Risk, Safety, and Ethical Concerns in Nanotechnology

Scrivener Publishing
100 Cummings Center, Suite 541J
Beverly, MA 01915-6106

Advances in Nanotechnology and Applications

This book series will focus on the properties and related applications of nanomaterials so as to have a clear fundamental picture as to why nanoparticles are being tried instead of traditional methods. Since nanotechnology encompasses various fields of science, each book will focus on one topic and will detail the basics to advanced science for the benefit of all levels of researchers.

Publishers at Scrivener
Martin Scrivener (martin@scrivenerpublishing.com)
Phillip Carmical (pcarmical@scrivenerpublishing.com)

Risk, Safety, and Ethical Concerns in Nanotechnology

Madhuri Sharon
Sharon Institute of Nanotechnology; Parishkar College of Global Excellence (Autonomous), Jaipur, Rajasthan, India

Scrivener Publishing

WILEY

This edition first published 2025 by John Wiley & Sons, Inc., 111 River Street, Hoboken, NJ 07030, USA and Scrivener Publishing LLC, 100 Cummings Center, Suite 541J, Beverly, MA 01915, USA

For more information about Scrivener publications please visit www.scrivenerpublishing.com.

Wiley Global Headquarters
111 River Street, Hoboken, NJ 07030, USA

For details of our global editorial offices, customer services, and more information about Wiley products visit us at www.wiley.com.

The manufacturer's authorized representative according to the EU General Product Safety Regulation is Wiley-VCH GmbH, Boschstr. 12, 69469 Weinheim, Germany, e-mail: Product_Safety@wiley.com.

AI disclosure statement: The author acknowledges the use of AI-based tools for minor language editing or grammatical correction. However, all intellectual contributions, analyses, and conclusions presented in these chapters are entirely the author's own.

Library of Congress Cataloging-in-Publication Data

ISBN 978-1-119-64038-7

Cover image: Generated with AI using Adobe Firefly
Cover design by Russell Richardson

Set in size of 11pt and Minion Pro by Manila Typesetting Company, Makati, Philippines

Printed in the USA

10 9 8 7 6 5 4 3 2 1

Dedicated to Prof. Maheshwar Sharon.

Contents

Foreword

In the annals of human progress, few fields have promised such profound and far-reaching implications as Nanotechnology. The ability to manipulate matter at the atomic and molecular levels heralds an age where medical treatments can target disease with unparalleled precision, materials can be engineered for exceptional strength or flexibility, and energy solutions can revolutionize sustainability. The dawn of the nanotechnology era has opened vast vistas of innovation, promising transformative advancements across diverse fields such as medicine, energy, and environmental science. Yet, alongside these extraordinary opportunities lie equally significant challenges—ethical dilemmas, unintended consequences, and risks that demand our utmost attention.

In the realm of nanotechnology, the invisible becomes tangible, the minute holds monumental potential, and the interplay between science and society reaches unprecedented depths. As we journey through this burgeoning field, it becomes imperative to balance progress with responsibility, innovation with caution, and exploration with ethical stewardship.

This book serves as a timely exploration of the intersection between risk and ethics in nanotechnology. It invites us to pause and consider: What responsibilities accompany the power to reshape the building blocks of nature? How do we weigh the potential benefits against the uncertainties and unintended effects of such transformative innovations? And, perhaps most critically, how can we ensure that this burgeoning field develops in a way that prioritizes equity, safety, and integrity?

Nanotechnology, by its nature, defies conventional boundaries—bridging physics, biology, chemistry, and engineering, and challenging traditional regulatory and ethical frameworks. The risks posed are as diverse as the field itself: environmental implications from nanomaterial accumulation, unforeseen health impacts, or even misuse in surveillance or weaponry. These risks underscore the importance of proactive governance, informed public discourse, and ethical foresight.

The author of this book Dr. Madhuri Sharon has meticulously examined the multifaceted risks associated with nanotechnology, from environmental impact and health hazards to societal implications and regulatory challenges. She delves into the ethical dilemmas that arise as we push the boundaries of what is scientifically possible, urging us to reflect on the moral dimensions of our technological pursuits.

In the pages ahead, you will encounter rigorous analysis, compelling case studies, and thought-provoking discussions that illuminate the ethical landscape of nanotechnology. This book is not merely a guide for scientists and engineers; it is a call to action for policymakers, educators, ethicists, and citizens to collaborate in shaping a responsible future for nanotechnology.

The pursuit of knowledge is a hallmark of humanity, but as history has shown, progress without prudence can lead to profound consequences. By addressing the risks and ethical considerations inherent in Nanotechnology, this work provides a compass for navigating the complexities of innovation in a rapidly changing world. The discourse on nanotechnology's risks and ethics is not merely academic—it is a call to action for policymakers, scientists, industry leaders, and the public. This book empowers its readers to make informed decisions, fostering a culture of transparency, accountability, and ethical integrity in the development and application of nanotechnologies

This book, *Risk, Safety, and Ethical Concerns in Nanotechnology*, serves as a comprehensive guide to navigating these complex landscapes. It is my hope that this book inspires both caution and optimism, urging readers to embrace the promise of nanotechnology while remaining vigilant about its perils. Together, let us ensure that the nanotechnological revolution is guided by wisdom as well as ambition, and that its legacy is one of progress that enhances, rather than endangers, the fabric of our global society.

As we stand on the threshold of a nanotechnology-driven future, let this book be our guide, illuminating the path forward with wisdom, caution, and a deep sense of responsibility. May it inspire thoughtful dialogue, rigorous scrutiny, and collaborative efforts to harness nanotechnology's potential while safeguarding the well-being of our planet and its inhabitants.

Professor Anand Ranganathan
Special Centre for Molecular Medicine
Jawaharlal Nehru University
Aruna Asaf Ali Marg, New Delhi, India

Preface

Life is either a daring adventure or nothing, – Hellen Keller

With this impetus, I have dared to write this volume. At the onset of penning down this book, I was in a confused state, uncertain whether I should even attempt to write it. But as the famous Urdu shayar Majrooh Sultanpuri wrote: *"Main akela hi chala tha janibe manzil, log aate gaye or karavan banata gaya."* Translation: *I set off alone towards my goal, but people came along and it began to turn into a caravan.*

I started with the idea of exploring the risks and impacts of nanoparticles, and along the way, hundreds of pieces of literature joined me—shaping this book into what it is. This work is a humble effort, aimed at researchers as well as commercial nanotechnologists, encouraging them to consider the potential risks associated with nanoparticle use.

This book is divided into two parts.

The First Part focuses on the various aspects of risks involved in researching, handling, and synthesizing nanoparticles. Before delving into these risks, an attempt is made to briefly introduce nanotechnology, nanoparticles, and their characteristics and properties. The interaction of nanoparticles with living systems and the environment is explored, as this interaction is central to understanding the health and environmental risks involved.

Particular attention is given to the lack of comprehensive data needed to fully characterize the health risks posed by various nanoparticles—specifically their toxic impacts, routes of exposure, and behavior in biological systems and ecosystems. As a relatively new field, there is limited knowledge of the long-term effects of nanoparticles on human health and the environment. This uncertainty presents a significant challenge. Therefore, the book proposes a coherent framework for analyzing available information to guide robust decisions by scientists and manufacturers dealing with risks and ethical concerns in nanotechnology.

This section also addresses responsible innovation and the intersection of definitions—what qualifies as "nano" and what constitutes "responsibility." Various methodologies for assessing health and environmental risks associated with nanoparticle-based products—such as cosmetics, food, defense, and theranostics—are presented. Routes of nanoparticle entry (inhalation, dermal absorption, biokinetics) and their relevance to risk are examined. Ecosystem risks are also discussed. This part is particularly aimed at professionals in corporate and regulatory sectors who must make decisions about nanoparticle safety in the absence of definitive evidence.

The Second Part of the book deals with ethics, governance, and policy —highlighting problems and concerns. It emphasizes ethical practices that should be followed by nanotechnologists, governments, national and international collaborators, and policymakers. It explores the health effects of nanoparticles, the technical issues in exposure assessment and management, and how current risk paradigms can be adapted or redefined to meet the ethical challenges posed by nanoparticle use.

A brief introduction is provided to various agencies and organizations focused on the ethical assessment of risks to health, society, and the environment. The recent COVID-19 pandemic—and the urgent race to develop therapies and vaccines—has served as a critical case study, sharpening the global focus on risk and ethics. Accordingly, two chapters are dedicated to concerns related to viruses and, in particular, COVID-19.

Finally, the book examines legal frameworks across various countries, international protocols, and the complex task of balancing the pros and cons of nanotechnology.

I sincerely thank Ms. Anuradha Pandey-Dubey for reviewing the text and helping to make figures. Additionally, I offer my gratitude to Martin Scrivener and the team at Scrivener Publishing for their efforts and guidance in the publication of this work.

Madhuri Sharon
August 10, 2025

Prologue

Basic Ingredients of Ethics and Human Legacy

We deal with ethics on a daily basis, from the time we are born to when we eventually pass on. From culture specific ideas like Sanskar, through which we try to teach the unspoken and unwritten rules of society to the next generation, to global institutions like the United Nations that aim to preserve ethical standards for the entire world. It is therefore not a surprise that when dealing with technology and science of any kind, working towards ethical science is a fundamental issue that not just scientists and engineers, but the entire society, has to collectively decide on. To do so, let us begin by considering the major questions regarding ethics.

(i) Why is ethics an existentially important question for our societies?
(ii) How have the instruments and beliefs used by societies to determine and enforce ethically good and bad behavior changed over human history?
(iii) How are modern governments and other institutions failing at ensuring ethical behavior?

In this book we will build our understanding of ethics and make an attempt to answer these questions. This exercise will help us contextualize ethics for nanotechnology that we address below.

Humans and Ethics

We consider humans to be distinct from the rest of the animals, mainly due to their ability to think ethically and act with complex planning. Ethics through shared empathy, in particular, is the distinctive line between an animal and a human. We may share several things in common regarding the body and behavior with animals, but humans have been developing differently from even the closest great ape species for over ten million years. Our mind has been diversified, widely and deeply, in cognitive and affective (emotional) perceptions of the world and of other minds around us. The emotional structure and functioning of humans have been altogether highly elevated themselves—no other members of a species can more deeply understand and predict the behavior of each other as we can. We humans have shed tears far more for each other than any other species. Though many animals have developed many superhuman cognitive senses, in the entire history of life on the Earth, ethical thinking is exclusively a very different psycho-asset, which distinguishes humans from animals, and which is the central life energy of human culture and experience.

This claim was verified through the research of C. Donald Batson, who says "we humans have the capacity to care for others for their sakes (altruism)" and "empathy-induced altruism is within the human repertoire." He further explains that empathy-induced altruism is a far more pervasive and powerful force in human affairs than has been recognized. Failure to appreciate its importance has handicapped attempts to understand why we humans act as we do and wherein our happiness lies. This failure has also handicapped efforts to promote better interpersonal relations and create a more caring, human society.

The struggle of all the past and present cultures can be viewed as the important story of human ethical conflict. The prominent stories of every religion and world epics, the innumerable philosophical schools of thought, the finest ideas in social sciences, the basic guidelines of national constitutions, the formal-informal teachings of gurus and teachers and the whispering of every parent into the ears of their cherished children carry the message of ethical thinking. In a way, all the cultures of the world have been cultivated by upholding the ethical values throughout the thousands of years of known human history.

The important question, then, is how do cultures decide on exactly which ethical ideas to follow and how to implement them. Everybody defines ethics as per their value structure, experience, moods and whims. Every person has their own say on the righteousness of any thought and action. Even established philosophies of the world, of the West and the East, are divided regarding the perception of ethics. The matter is further complicated by the fact that ethical conclusions can change with situations and contexts.

The Western thinkers, from Socrates to Kant, Marx and Derrida and Eastern philosophies depicted in Vedas, Upanishads, Ramayana and Mahabharata epics, Jainism, Buddhism, Zoroastrianism, Islamism, and Sikhism, all commonly agree that the aim of ethical life is primarily to control animal-like selfish thinking and behavior and to dedicate oneself to the welfare of the "others" — local society and global humanity. All the cultures unanimously agree that the aim of every human's life is enjoyment

(Anand), which is the culmination of feeling accomplishment, and that it is possible only through one's thoughts and deeds dedicated to the "welfare" of the "other." For example, Dalai Lama, argues that the most important thing for us is to conduct our lives with love and compassion for others, and our societies need to develop a stronger notion of universal responsibility and acknowledge our inherent interdependency.

Failure to act ethically can and does create conflicts. On the other hand, welfare-oriented deeds liquidate the "self" into the "other," and thus gives us the feeling of accomplishment and enjoyment. Of course, this perception may seem fantastical in this modern world, but if everyone thinks and acts for the welfare of the "other," the perpetual agony of mankind—manifesting in the form of a fight between two family members up to the heinous world wars that devastated all of humanity—will be ameliorated. The ultimate hope is for a positive movement by all the cultures where every human being is considered an equal valuable member of the world culture (Vasudhaiva Kutumbakam — meaning that if there is any concept of family, it includes the whole earth; the concept of a family cannot be smaller than that). Thus, every member of this family — humans, animals, birds and other creatures — are cared for and honored for everyone's optimum development.

God and Ethics

For the majority of cultures throughout the history of religion, a rough system of beliefs and practices based on a monotheistic or a pantheon of gods would serve as the instrument to decide and enforce ethics in society. In this way, God, or god(s), is a personified myth of collective virtues. The "death of God," as Nietzsche would put it, came about with the rapid emergence of science and technology, evaporating the power that religion alone would determine. It can be argued that this "death of God" was in a way the "death of ethics," freeing humans from most moral obligations and making them "liberal."

Since the eighteenth to the first quarter of the twenty-first century, the concept of ethics has been unethically defined and replaced by many illuminating post-medieval concepts. These concepts tend to be fairly attractive to us—liberty, individual freedom, personal privacy, individual property, business rights, democracy, fundamental rights, human rights, capitalist economy, corporate economy, etc. The impact of social struggles like the American Revolution (1775-1783) and the French Revolution (1789-1799) spread roaringly throughout the world with a trio of values—Liberty, Equality and Fraternity; but Liberty, in isolation, mischievously damaged the concept of Equality and especially Fraternity; and ultimately today, they have resulted in countries and states that are best described as crony capitalisms and electoral autocracies. "Others, noting the evolution of the market society, understood that the liberty of some would entail the alienation of others, and they sought Equality" (Jacques Attali, "Fraternity and Utopia" 2000).

Liberty, Science and Technology, Capitalism and Ethics

Now, with the rise of capitalism as the popular form of economic framework, competitive profits and money have been accepted as the key inspirations of any innovation and development. Due to the development of science and technology, we now have production, distribution, and consumption on a grander scale than ever before; however, this has led to an accumulation and centralization of profit and power in the hands of an increasing minority of the world. "We are aggressively pursuing the promises of these new technologies within the now-unchallenged system of global capitalism and its manifold incentives and competitive pressures" (Bill Joy, *Wired*, 2000).

The concepts of "liberty" and "freedom" are used by the governments of capitalist states to legitimize the profit orientation and its centralization. They have transformed the medieval decentralized and distributed economy into a centralized capitalistic economy; and open and co-operative economic participation into a patent-based monopolistic economy. Unfortunately, by ignoring the ethics of the welfare of deprived people we have transformed our "democratic" politics into a commercialized money-power game. With a great deal of confusion and derailment, most states have become "ethics-neutral business center" with increased exploitation and oppression of the weaker classes, hiding it through unethical legitimations

Consequently, now we have more inequality amongst the population, more disparity in economic and political participation, more political tyranny in even so-called democratic states and more antagonism at every level. This is apparent as the measures of economic equality and democratic integrity of states indicate (Economist Intelligence, 2022). The status quo in 2024 has exposed the poverty in ethical forethought and practice at the global level—the wars between Russia and Ukraine, Israel and Palestine, the incompetence of the UN in resolving the current conflicts; the ignorance at world forums regarding the deaths and agony of children, women and men; the dubious desires of the rich countries for political and economic benefits that increase the miseries of underdeveloped countries.

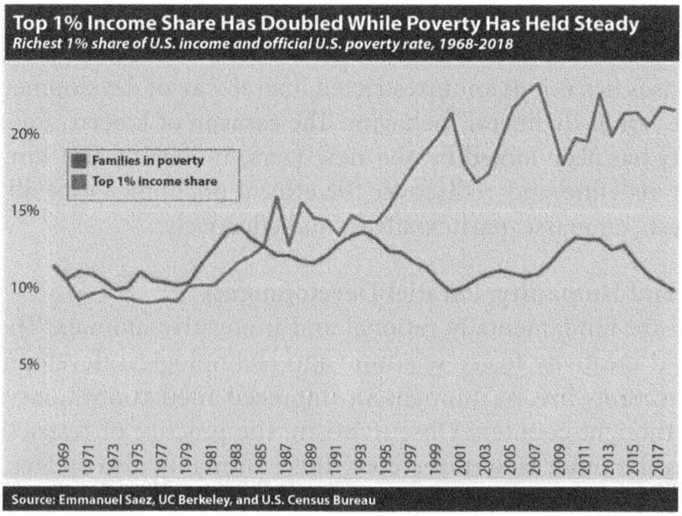

The whole world is now slowly waking up to the man-made global imbalances in the environment—global warming and large-scale pollution threaten the existence of every coastal city and the intensity and frequency of natural disasters and higher rates of horrible diseases like cancer are increasing—stemming from the goal of seizing maximal profits. Besides existential threats, the quality of life of the most vulnerable and powerless sections of society is also under threat due to increased joblessness, exploitation of labor in sweatshop-like jobs (often protected by political support) and reduced upward social and economic mobility, all of which can collectively be seen as a squeeze on political and economic liberty. This is almost a paradox—the majority population of every state is the most crucial in the development process, but they are the most neglected. They are the ones most deprived of opportunities in health, education, employment, and political rights despite an explosion of collective human economic power. The whole world is becoming more and more unethical regarding the political rights and participation of the common man. A few rich people's desires have polluted the urban air inhaled by the lungs of the poor people, while the rich themselves can escape to homes surrounded by the clearer countryside air. It is becoming increasingly clear that the very "liberty" of an individual or an industrial company has polluted the entire world's environment, which poses a serious danger to the housing, food, socializing, and economic and political participation of the general masses, thus damaging the "liberty" of the rest of the people. This is a blatant aggression by the lavishly "liberated" rich industrialists, supported by powerful "liberated" politicians, built on the "liberty"

and even the "existence" of the poor people of not just the same country, but also of another country. Thus, the solution to all these mismanaged global crises lies not in an unrestricted liberal way of development, but in an active "ethically liberal" behavior. The caravan of Liberty, Equality and Fraternity has been looted by the new tzars, monarchs and kings. Thus, we must redefine and rediscover the ethical phenomena for all of mankind afresh, en masse, participatively and effectively.

Science and Humanity: Parallel Developments

Humans are fundamentally rational and innovative animals. Their rationality and creativity foster scientific and technological development and the innovations involve humans in unprecedented continuous research. Human thought counters Obscurantism, the practice of restricting curiosity. On the other side of the coin, their animality also creates destruction, which is curtailed through rationality-based deterrence. Sure, there have been wars throughout human history, but the times of peace have, in general, been more prevalent. To understand the power of deterrence, one need not look beyond the fact that humans possess thousands of atomic weapons all around the world, but we haven't used them since 1945. On the other hand, development of nuclear power has offered a fairly clean and safe source of energy.

Our scientific and technological developments have been expanding not just vertically but also horizontally. In every prominent industry and at every major university, scholars are establishing their identities by exploring innovative ideas and techniques. The widespread desire for education and increased exposure to scientific ideas and technological devices are regularly creating opportunities for inquisitive thinking and practicing. A series of revolutions in science and technology are accompanying social and humanistic thoughts and behavior.

Human society is like a global ocean, where all the rivers and streams of scientific inventions and cultural thoughts are being merged all the time. There is not a single drop of water in isolation in nature; it shares its own humidity with every touch of the air. In both these ways scientific developments—Galileo (1638) & Newton (1704) studied the nature of light; Boyle (1688) researched the compressibility of gas; iii) James Watt (1764) improved the steam engine; Halley (1758) predicted the orbit of comets; Cavendish (1776) discovered hydrogen; Laplace (1796) theorized the origin of the solar system; Cavendish (1798) measured Earth's mass—changed the technology and the physical production world, differently; and it also revolutionized the medieval perception of socio-political-economic issues altogether in the form of the American Revolution (1775-1783) and French

Revolution (1789-1799) in the same period of history. Reciprocatively, the American and French Revolutions also enhanced the Scientific Revolution; impacting each other and changing every aspect of life primarily in the USA and Europe; and sooner or later throughout the world. We fought the First World War in 1914-19 and we also formed the League of Nations in 1920; we used atom bombs at Hiroshima and Nagasaki in August 1945, and we also formed the UN, an organization to maintain the peace, in October 1945, just two months later.

A scientific revolution explores the process of nature at a fundamental level, it acknowledges the hidden relationships of the elements of nature, and at the same time, reciprocally it also reorganizes the new production system, new economic organization, new political order, and consequently also new ethical fiber. These are all related, integrated, simultaneous happenings; stimulating and accelerating each other and creating and recreating a new order of the world. As nanoscience and nanotechnology is a new revolution in our scientific progress pyramid, it must be supported by a new set of ethics in our political-economic system, along with a new order in every nano-production and its consumption process.

Science and Humanities

A consequence of the increased importance of science and technology is that the popularity of a liberal art education has been replaced by a technical education. Thus, social sciences have been set aside after being deemed unnecessary, leaving ethical forethought wanting in science and tech. Unfortunately, all the educational institutions, cultural forums, and social organizations also accepted the value-free, ethics-neutral way of life. Almost all the political structures accepted that unethical actions—inequality, exploitation, state violence, wars, selfishness, dishonesty, falsehood, fakeness, pretentions, fraud, corruption—are natural and expected from individuals and companies of the state itself. To earn money, through any sort of means, became the legitimate right in the minds of general masses.

Dr. Raghava Prakash
Director, Parishkar College of Global Excellence
Jaipur, India

Part I

NANOTECHNOLOGY AND RISKS

1

Understanding Nanoparticles and Their Probable Risks

You cannot swim for new horizons until you dare to lose sight of the shore.

William Faulkner

Abstract
This chapter introduces the first part of the book, which focuses on the risks posed by emerging nanotechnology. As industries race to commercialize nanomaterials, safety concerns—especially long-term health and environmental effects—remain underexplored. The chapter outlines the unique properties of nanomaterials and their growing presence in fields like healthcare, electronics, agriculture, and defense. It emphasizes the urgent need for regulation, risk assessment, and research on exposure, dosage, and solubility of nanoparticles, given their ability to enter the body and disrupt cellular processes.

Keywords: Nanoparticle-fabrication, magnetic-properties, electrical-properties, thermal-properties, chemical-properties, mechanical-properties, optical-properties, consumer-products

1.1 What is Nanotechnology?

Before understanding the risks associated with nanoparticles, we must know what nanotechnology is.

Nanotechnology—the word alone is enough to make anyone scratch their head. Science of the small? How does that work? Why use nanotechnology? Why is it assumed that smaller is better? These are all questions that arise when talking about nanotechnology. As this field becomes more popular

Madhuri Sharon. Risk, Safety, and Ethical Concerns in Nanotechnology, (3–18) © 2025 Scrivener Publishing LLC

Figure 1.1 (Left to right) Propounder of Nanotechnology Nobel Laureate Richrd Feynman; Prof. Norio Taniguchi; and Dr. Kim Eric Drexler.

and useful in science, our understanding of it seems to be getting more and more complex; and every time a solution is found, 20 other questions appear. Let us start with what nanotechnology is.

The idea of nanotechnology was first introduced (Figure 1.1) by Richard Feynman in 1959 [1] in a talk entitled "There's Plenty of Room at the Bottom." Later, further experimentation based on this idea led to the term "Nanotechnology" being coined by Norio Taniguchi in 1974 [2] (he must have taken the word Nano from the Greek word which means extremely small or dwarf). The potential of Feynman's idea was exhibited in the mid-80s by Eric Drexler [3] in his book *Engines of Creation* in which the potential of molecular nanotechnology is explained.

A *nanometer* unit is 10^{-9} m. For nanotechnological purposes, the size of nanoparticles is accepted to be from 100 nm down to the size of atoms, because it is at this scale that the properties of materials can be very different from those at a larger or bulk scale. Nanoparticles are neither visible to the naked eye (which cannot see things smaller than 0.2 mm) nor by the light microscope under which things smaller than 0.2 um cannot be seen.

In a nutshell, nanotechnology involves synthesis design, characterization, production, and application of structures, devices, and systems by controlling shape and size at the nanometer scale as well as application of nanoparticles.

The U.S. National Nanotechnology Initiative defines nanotechnology as "The science, engineering, and technology related to the understanding and control of matter at the length scale of approximately one to hundred nanometers. However, nanotechnology is not merely working with matter at the nanoscale, but also research and development of materials, devices, and systems that have novel properties and functions due to their nanoscale dimensions or components."

Nanoparticles are fundamental units of nanotechnology and are used as raw materials for constructing nanostructures, nanomaterials, nanomachines, and nanodevices. Nanoparticles may be defined as "A particle having one or more dimensions of 100 nm or less."

Every month, thousands of publications appear that cover disciplines, such as chemistry, physics, life sciences, medicine, and engineering, informing the current status of the development of nanotechnology and its applications.

1.2 Unique Properties of Nanoparticles that Differentiate Them from Their Bulk Counterpart

Man has always been intrigued by the behavior of various mechanisms at the sub-atomic level. The properties of physics, chemistry, and biology are altered at this level and many times the laws of physics and chemistry are altered to a great extent, leading to materials having properties that can be exploited for a plethora of commercial and fundamental applications. In this dynamic legacy of atoms and molecules, we step onto a very different platform, where the classical laws of physics used to understand the behavior of material become deceptive. Hence, a very sharp weapon of fundamental physics called quantum mechanics dominates this ever-changing dynasty of material properties. The change in the properties of materials when they combine in many ways became clearer when, due to the invention of instruments, we were able to peek into the world of atoms and molecules. This congregation of molecular arrangements can generate machinery at a very small scale and give birth to one of the most potential subjects of the century, called nanotechnology.

The classical laws of physics and chemistry do not readily apply at this very small scale for two reasons. Firstly, the electronic and optical properties of very small particles can be very different from their larger particle. Secondly, the ratio of surface area to volume becomes much higher, and since the surface atoms are generally most reactive, the properties of a material changes in unexpected ways.

Nanoscience is not just the science of small, but the science in which materials with small dimensions show new physical phenomena, collectively called quantum effects, which are size-dependent and dramatically different from the properties of macro materials.

At the nanoscale, there is a vast increase in the ratio of surface area to volume. Due to this, materials at nanoscale show very different reactive properties as compared to what they exhibit on a microscale; and it is still very difficult to predict the physical and chemical properties of particles of such a very small size. For instance, opaque substances become transparent (copper), inert material becomes catalysts (platinum), stable material turns combustible (aluminum), solids turn into liquids at room temperature (gold), insulators become conductors (silicon). The principal parameters that affect the properties of nanoparticles are their shape (including aspect ratios where appropriate), size, the morphological substructure of the substance, surface characteristics, and inner structure. Nanoparticles can exist as aerosols (solids or liquids in the air), suspensions (solids in liquids), or as emulsions (liquids in liquids).

The composition of a specific nanoparticle can be very complex, depending on what interactions it has had with other chemicals or particles and on its lifetime. The chemical processes taking place on the surfaces of nanoparticles are also very complicated and remain largely unknown. Nanoparticles have different ways of interacting with each other. They can remain free or group together depending on the attractive or repulsive interaction forces between them. These interactions remain difficult to characterize.

Some unique physical and chemical properties of nanoparticles are listed below:

1.2.1 Magnetic Properties

Magnetic properties of nanostructured materials are distinctly different from that of bulk materials. Ferromagnetic particles become unstable when the particle size reduces below a certain size as the increase in surface energy provides sufficient energy for domains to spontaneously switch polarization directions and become paramagnetic. But this transformed paramagnetism behaves differently from the conventional paramagnetism and thus is referred to as superparamagnetism. In other words, ferromagnetism of bulk materials disappears and gets transferred to superpara magnetism in the nanoscale due to the high surface energy.

1.2.2 Electrical Properties

Nanoparticles are good conductors of electricity, which make the electrical devices faster with lower power consumption and with reduced imperfections. Some metals become semiconductors as the size decreases because of quantum confinement, e.g., quantum dots.

1.2.3 Thermal Properties

Metal and semiconductor nanoparticles are found to have a significantly lower melting point or phase transition temperature as compared to their bulk counterparts. The lowering of the melting points is observed when the particle size is >100 nm and is attributed to an increase in surface energy with a reduction of size. The decrease in the phase transition temperature can be ascribed to the changes in the ratio of surface energy to volume energy as a function of size.

1.2.4 Mechanical Properties

At the nanoscale, many of the mechanical properties of the materials are modified such as hardness, fracture toughness, scratch resistance, and fatigue strength. Energy dissipation, mechanical coupling within arrays of components, and mechanical nonlinearities are influenced by structuring components at the nanometer scale. The mechanical properties of nanomaterials increase with the decrease in size. Most of the studies have been focused on the mechanical properties of one-dimensional structures such as nanowires. The enhanced mechanical strength of nanowires or nanorods is ascribed to the high internal perfection of the nanowires. Generally, imperfections such as dislocations, microtwins, and impurities in crystals are highly energetic and should be eliminated from the perfect crystal structures. The smaller the cross section of nanowires, the less is the probability of finding any imperfections in them as the nanoscale dimension makes the elimination of such imperfections possible.

1.2.5 Chemical/Catalytic Properties

The huge surface-to-volume ratio of nanoparticles increases its chemical activities and therefore allows nanomaterials to become an efficient catalyst. Moreover, due to the properties of interacting with biomolecules, nanoparticles exhibit toxicity, reactivity, and also enhanced cellular uptake.

1.2.6 Optical Property

For the last few decades, metallic nanoparticles have fascinated researchers due to their colorful colloidal solutions. Mie was the first to explain the red color of gold nanoparticles in 1908 by solving Maxwell's equation for an electromagnetic light wave interacting with small metallic spheres. The color exhibited by metallic nanoparticles is due to the coherent excitation of all the "free" electrons within the conduction band, leading to an in-phase oscillation, which is known as surface plasmon resonance. Thus, the color of metallic nanoparticles may change with their size due to surface plasmon resonance. The unique optical property of nanomaterials may also be due to the quantum size effect, which arises primarily because of the confinement of electrons within particles of dimension smaller than the bulk electron delocalization length. This effect is more pronounced for semiconductor nanoparticles, where the bandgap increases with decreasing size. The same quantum size effect is also shown by metal nanoparticles when the particle size is >2 nm. Depending on their chemical composition, their optical and electronic properties have generated much attention in studies of properties of particles approaching molecular dimensions.

The properties of nanoparticles, viz., physical, chemical, and biological can be manipulated for the desired objectives. Nanostructures and nanomaterials possess a large fraction of surface atoms per unit volume. The total surface energy increases with the overall surface area, which is in turn strongly dependent on the dimension of the material.

1.3 Fabrication of Nanoparticles

The synthesis of nanomaterials is also called *nanofabrication*. There are two approaches (Figure 1.2 & Table 1.1) involved in the synthesis and fabrication of nanostructures:

(i) *Bottom-up approach* involves assembling single atoms and molecules into larger nanostructures. The bottom-up approach refers to the buildup of material from the bottom, atom-by-atom, molecule-by-molecule, or cluster-by-cluster. This method creates identical structures with atomic precision. Moreover, by arranging molecules to form complex structures, new and useful properties have also been accomplished by this method. The advantage this approach has is that it creates nanostructures with fewer defects while a much more homogeneous chemical composition and better short- and long-range ordering can be obtained. However, it is difficult to make a structure large enough and of desired quality. This approach has the potential of producing less waste and therefore is more economical.

(ii) *Top-down approach* involves breaking down large pieces of material to generate nanostructures or nanoparticles

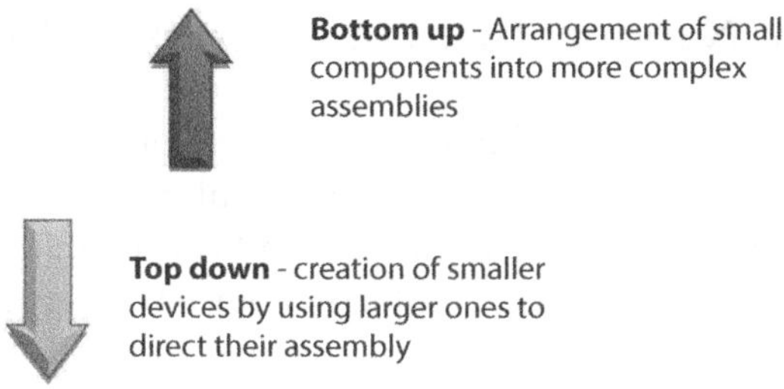

Figure 1.2 Approaches for assembling nanoparticles.

Table 1.1 Examples of various synthesis methods using top-down and bottom-up approaches.

Top-down approach using	Bottom-up approach using
Thermal method	The chemical vapor deposition method
High energy method	Electrodeposition method
Lithographic method	Liquid phase method
Templet method	Seed mediated growth method

of desired size and shape by employing ultrafine grinders, lasers, and vaporizations followed by cooling. The main challenge for the top-down approach is the creation of increasingly small structures with accuracy. Moreover, the top-down approach usually introduces internal stress and surface defects.

These techniques are an extension of those that have been used for producing micron-sized particles. This method is often used for making interconnected and integrated structures, for example, in electronic circuits.

The nanoparticles synthesized by various methods are in the form of colloids, clusters, powders, tubes, rods, wires, and thin films. Apart from various physical and chemical methods of fabricating nanoparticles; there are biological and hybrid techniques available for the synthesis of various nanomaterials. As alternative biological methods or green technology are considered to be safe and sound, the use of biological methods has many advantages over traditional methods. The critical factors in developing a bio-directed self-assembly approach are identifying the appropriate compatibilities and combinations of biological-inorganic material. Biogenic nanoparticles are not supposed to use harmful chemicals.

1.4 Uses of Nanoparticles in Consumer Products

We are living in a time of technological innovation. Nanotechnology, the science of manipulating material at the molecular scale, is a technological development that is moving fast into use and commercialization in many sectors. Over the last few years, nanotechnology has emerged as the science of nanoparticles whose use in many consumer products is very popular. The potential uses and benefits of nanotechnology are enormous. The burgeoning nanoscience field has promised everything from mundane things like better paints, self-cleaning windows, cosmetics to the bizarre tiny submarines that will glide through our veins destroying pathogens and parasites [4].

Today's nanotechnology harnesses current progress in chemistry, physics, materials science, and biotechnology to create novel materials that have unique properties because their structures are determined on the nanometer scale. Nanotechnology has already been embraced by industrial sectors, such as the information and communications sectors, but is also used in food technology, energy technology, as well as in some medical products and medicines.

Already several nanoscale structures are developed and in practical use in areas related to materials science, biology, and medicine, especially in drug designing; targeted drug delivery, tissue engineering, and imaging are also being developed for analytical and instrumental applications. A wide variety of nanoscale materials and coatings are already in use in consumer products such as cosmetics and sunscreens, fibers and textiles, dyes, and paints. Moreover, data storage devices based on nanostructures provide smaller, faster, and lower consumption systems; and new types of microscopes that can produce images of atomic and molecular processes at the surface.

Some of the materials using nanotechnology and nanoparticles that are in the consumer market today are:

- Antimicrobial powders to kill pathogens and contaminants;
- Cosmetics such as sunscreen, nano-liposome-based skin-care products;
- House-cleaning products such as window-washing sprays;
- Lab-on-a-chip devices, biosensors, medical imaging, prostheses and implants, drug delivery devices, etc., for biomedical applications;
- Lighter and stronger ceramics;
- Nanocatalysts;

- Nanoelectronics used in information and communication technology;
- Non-permeable membranes for food packaging;
- Paints, varnishes, and other coatings;
- Remote sensing and tracking devices;
- Sensors, seed improvement, and nano-fertilizer to support agriculture and the environment;
- Solar cells, fuel cells, batteries;
- Various applications in military technology [5].

Apart from the materials that have already found their way into consumer products, others are being intensively researched for solutions to humanity's greatest problems—diseases, clean energy, clean water, etc. Other work is aimed at developing a roadmap for a productive nanosystem, in which a path is sought from today's nanotechnology capabilities to advanced future systems in which molecular tools will build useful materials, devices, and complex systems to atomic precision. The products of advanced nanotechnology that will become available in coming decades promise even more revolutionary applications than the products of current and near-term nanotechnology are providing in the hope of solving critical challenges facing humanity such as:

- Enabling space development;
- Healing and preserving the environment;
- Improving health and longevity;
- Making information technology available to all;
- Maximizing productivity of agriculture;
- Providing renewable clean energy;
- Supplying clean water globally.

1.5 Potential Risks Due to Harmful Effects of Nanoparticles

To know the potentially harmful effects of nanoparticles, we have to know whether and how nanoparticles interact with living organisms, what are the health implications of nanoparticles, do they affect the environment, and above all how to assess the effect of nanoparticles. In this chapter, as well as in this part of the book, we will be addressing this.

Though nanotechnology and nanomaterials offer new opportunities for umpteen applications and even reduction of environmental pollution, we cannot overlook the possible dark side of nanotechnology as it may present new health and environmental risks. Humans have to develop mechanisms of protection against various synthetic nanoparticles of different sizes and their specific characteristics. Chances are that the normal human defense mechanisms, such as immune and inflammatory systems, may not be able to respond adequately to the nanoparticles. Moreover, the dispersal and persistence of nanoparticles in the environment will also have an impact on the environment. For assessing the health risks, two types of nanostructures should be considered: (i) free nanoparticles, and (ii) nanostructure that are an integral part of larger materials such as coatings composed of nanomaterials and other ingredients. Free nanoparticles will pose more interaction with the living system and they pose a greater risk for health or the environment than the nanoparticles fixed to other compounds.

1.5.1 Risk Due to Interaction of Nanoparticles with Living Organisms

Studies dealing with the interaction of nanoparticles leading to toxicity have recently attracted attention. However, they are still scarce. The key factors that affect the interaction of nanoparticles with living structures are:

(i) *Dimensions of Nanoparticles:* Some nanoparticles, have the same dimensions as biological molecules such as proteins. Nanoparticles that are no bigger than a few nm can reach inside biomolecules, which is not possible for larger nanoparticles. Moreover, nanoparticles can cross cell membranes. If inhaled, the nanoparticles can reach through the blood to other sites such as the liver, heart, or blood cells.

(ii) *Surface Characteristics of Nanoparticles:* Nanoparticles may immediately get adsorbed onto the surface of some of the large molecules of living systems they encounter, as they enter the tissues and fluids of the body. Nanoparticles have the unique ability to stick to the molecules of the surface of the living system, due to surface characteristics. This property is useful and relevant for targeted drug delivery using nanoparticles.

(iii) *Dose and Solubility of Nanoparticles:* These are also two of the key factors that help them spread within the body.

Nanoparticles that dissolve easily affect the living organisms in the same way as the effects of the chemical they are made of. But some nanoparticles do not degrade or dissolve readily; hence, they accumulate in biological systems and persist for a long time, which makes such nanoparticles of particular concern.

(iv) *Size of Nanoparticles:* The impact of these factors on the interaction with the living system and the response of living organisms to such nanoparticles are not thoroughly understood. There is a need to understand and categorize the toxicity of nanoparticles. Since nanoparticles of any material are very small in size, they have a much greater surface-to-volume ratio; therefore, relatively more molecules of the chemical are present on the surface. This may be one of the reasons why nanoparticles are generally more toxic than larger particles of the same composition. In addition to being able to cross cell membranes, the nanosized particles reach the blood and various organs easier than larger particles of that same material.

(v) *Shape of Nanoparticles:* Though the effects of nanoparticles on health and the environment have not been evidenced in detail, these are also likely to depend on their shape. For example, CNTs are highly toxic as they produce harmful effects by an entirely new mechanism, different from the normal model of toxic dust.

(vi) *Chemical Composition and Surface Characteristics of Nanoparticles:* These have been found to affect the toxicity of nanoparticles. The composition of any chemicals adsorbed onto the surfaces of nanoparticles also affects the toxicity. However, successful trials have revealed that the modification of surfaces of nanoparticles can make them less harmful to health.

1.5.2 Short- and Long-Term Risks

The first step of chemical safety is hazard assessment, collection, and evaluation of available and relevant information, identifying the presence or absence of intrinsic hazardous properties of the substance. The second step is a comprehensive human hazard assessment that needs information regarding the fate of nanoparticles in the body (e.g., toxico-kinetics of absorption, distribution, metabolism, and excretion; and acute toxicity, irritation and corrosivity, sensitization, repeated dose toxicity, mutagenicity, carcinogenicity, and reproductive toxicity, etc.); and/or a comprehensive environmental hazard assessment, on the impact of nanoparticles on freshwater and marine water (acute and chronic toxicity to aquatic organisms), sediments and soil (toxicity to benthic or sediment-dwelling) organisms, sewage treatment plant and air (toxicity to sewage treatment plant (STP) microorganisms and other relevant environmental properties such as aquatic and terrestrial bioconcentration and bioaccumulation).

This hazardous information will identify the dose descriptor of the substance, which represents a level of exposure above which humans should not be exposed, and the predicted no-effect-concentration (PNEC), i.e., concentration of a chemical, which has unacceptable effects on the ecosystem and its organisms that will not occur during the long-term or short-term exposure. The PNEC is typically given in mg/L or mg/kg.

Guidance on available methods for risk assessment of nanomaterials has been prepared in 2015 by LIFE REACH nano, which has developed a web-based REACH Toolkit to support the chemical safety assessment of nanomaterials.

1.6 Need for Measurement of Exposure to Nanoparticles

In the absence of specific parameters to be considered for measurement of exposure to nanoparticles by an organism, the traditional risk assessment paradigm also holds for nanomaterials [6–9]. The basic components for assessing the risk of chemicals are hazard and exposure, including dose-response estimation, risk characterization, and accounting for uncertainty in the overall assessment. However, many of the tools, test protocols, and guidelines for the assessment of physicochemical properties, fate, exposures, and effects for conventional chemicals, need modifications if they are to be applied to the evaluation of nanomaterials [6]. Because many nanomaterials show particle-specific behavior, they require some specific considerations such as

(i) Different nanoforms can display different behavior both in their fate/toxicokinetics and hazard, and thus in risk. Hence, there is a demand to consider this in those dossiers for regulatory acceptance that comprise more than one nanoform. To consider this regulatory gap, different nanoforms should be addressed, which requires further research and policy considerations. Recently ECHA [10] has given some guidelines on the term "nanoform."

(ii) Life cycle and exposure are another issue raised by many scientists, because the physicochemical properties of nanomaterials may change during their life cycle [11, 12], which may form more hazardous nanomaterials during the life cycle; for example, due to the integration of a coating or photoactivation. It is imperative to consider this in general terms in risk assessment frameworks [13–15]. Other problems that may be encountered are a quantitative measurement of concentrations of nanomaterials such as degree of agglomeration, size distribution, etc., to be measured in complex matrices. This needs the development of computational models for estimating exposures and changes in physicochemical properties during the life cycle.

(iii) Delivered dose of nanomaterial is another criterion that needs attention because it tends to agglomerate, which can have a serious impact on the application of assessment factors [16, 17], especially if the internalized dose is still technically challenging. It requires correlating the *in-vitro* and *in-vivo* data to find ways forward for alternative testing.

(iv) Bioaccumulation and ecotoxicity is an important consideration as their elimination from tissues can be very slow and may take years, especially for hydrophobic non-nano materials. But the basic mechanisms of uptake and elimination differ in nanomaterials and the behavior of nanomaterials cannot be described by chemical equilibrium. Hund-Rinke *et al.* [18] have suggested that for environmental risk assessment of nanomaterials, a pragmatic screening procedure will be needed to determine the nanomaterial concentration in suitable test organisms; whereas for human health risk assessment of nanomaterials, lysosomal fluid may give information on the potential for accumulation.

(v) Assessment factors will vary due to the interspecies and intraspecies differences. The OECD [19, 20] indicates that there are insufficient data to verify if the present standard assessment factors (AFs) of 10 for both interspecies and intraspecies differences apply to nanomaterials.

Indeed, insufficient studies are available that can be directly compared to each other to allow assessment of the variation between studies.

Relevance of exposure, hazard assessment of nanomaterials, such as life cycle, bioaccumulation, and delivered dose, will be discussed in other chapters of this book along with risk assessment frameworks.

1.7 Conclusion

Whenever a commercially applicable branch of science is developed; many industries mushroom to take the early financial advantage. Many nanomaterial-based products are coming on the market. Being a new branch of science, many risks and disadvantages, especially in the long term, have not been assessed. This chapter has given a very basic introduction to nanotechnology and discussed the need to take into consideration the specific regulations, possible hazards, and envisaged risks, which will be discussed further in subsequent chapters.

References

1. Feynman, R., Talk at the annual meeting of American Physical Society 20 December 1959. *Caltech Eng. Sci.*, 23:5, 22–36, 1960, 1959, http://www.zyvex.com/nanotech/feynman.html.
2. Taniguchi, N., On the Basic Concept of Nanotechnology. *Proceedings of the International Conference on Production Engineering*, vol. 18-23, 1974.
3. Drexler, K.E., *Engines of Creation*, Anchor Books, New York, 1986.
4. Salata, O.V., Applications of nanoparticles in biology and medicine. *J. Nanobiotechnol.*, 2004.
5. Sharon, M., Silvestre Lopez Rodriguez, A., Sifuentes Gallardo, P., Sharon, C., *Nanotechnology in the Defense Industry: Advances, Innovation, and Practical Applications*, Wiley Scrivener, USA, 2019.
6. Sayre, P., Steinhäuser, K., van Teunenbroek, T., Methods and data for regulatory risk assessment of nanomaterials: questions for an expert consultation. *NanoImpact*, 2017.
7. SCENIHR, Opinion on the Appropriateness of Existing Methodologies to Assess the Potential Risks Associated with Engineered and Adventitious Products of Nanotechnologies. *European Commission*, 2005, Available: http://ec.europa.eu/health/ph_risk/committees/04_scenihr/docs/scenihr_o_003b.pdf, Accessed date: 11 September 2016.
8. SCENIHR, The Appropriateness of the Risk Assessment Methodology in Accordance with the Technical Guidance Documents for New and Existing Substances for Assessing the Risks of Nanomaterials. *European Commission Health & Consumer Protection Directorate- General*, 2007,

Available: http://ec.europa.eu/health/ph_ risk/committees/04_scenihr/docs/scenihr_o_010.pdf, Accessed date: 11 September 2016.

9. SCENIHR, Opinion on: Risk Assessment of Products of Nanotechnologies. *European Commission*, 2009, Available: http://ec.europa.eu/health/archive/ph_risk/ committees/04_scenihr/docs/scenihr_o_023.pdf.
10. ECHA, Appendix 4: Recommendations for Nanomaterials Applicable to the Guidance on Registration. Draft (Public) Version 1.0 January 2017, 2017a, https://echa.europa.eu/documents/10162/23047722/appendix_4_nano_registration_committees_ en.pdf/1abb12d1-88a2-b386-0907-c67d05105378.
11. OECD, Important Issues on Risk Assessment of Manufactured Nanomaterials, Series on the Safety of Manufactured Nanomaterials No. 33, 2012a, http://www.oecd.org/officialdocuments/publicdisplaydocumentpdf/?cote=env/jm/mono(2012)8& doclanguage=en.
12. Gottardo, S., Alessandrelli, M., Amenta, V., Atluri, R., Barberio, G., Bekker, C., Bergonzo, P., Bleeker, E., Booth, A.M., Borges, T., Buttol, P., Carlander, D., Castelli, S., Chevillard, S., Clavaguera, S., Dekkers, S., Delpivo, C., Di Prospero Fanghella, P., Dusinska, M., Einola, J., Ekokoski, E., Fito, C., Gouveia, H., Grall, R., Hoehener, K., Jantunen, P., Johanson, G., Laux, P., Lehmann, H.C., Leinonen, R., Mech, A., Micheletti, C., Noorlander, C., Olof-Mattsson, M., Oomen, A., Quiros Pesudo, L., Letizia Polci, M., Prina-Mello, A., Rasmussen, K., Rauscher, H., Sanchez Jimenez, A., Riego Sintes, J., Scalbi, S., Sergent, J.-A., Stockmann-Juvala, H., Simko, M., Sips, A., Suarez, B., Sumrein, A., van Tongeren, M., Vázquez-Campos, S., Vital, N., Walser, T., Wijnhoven, S., Crutzen, H., NANoREG Framework for the Safety Assessment of Nanomaterials, EUR 28550 EN, 2017. http://dx.doi.org/10.2760/245972.
13. Oomen, A.G., Bleeker, E.A., Bos, P.M., van Broekhuizen, F., Gottardo, S., Groenewold, M., Hristozov, D., Hund-Rinke, K., Irfan, M.A., Marcomini, A., Peijnenburg, W.J., Rasmussen, K., Jimenez, A.S., Scott-Fordsmand, J.J., van Tongeren, M., Wiench, K., Wohlleben, W., Landsiedel, R., Grouping and read-across approaches for risk assessment of nanomaterials. *Int. J. Environ. Res. Public Health*, 12, 13415–13434, 2015.
14. Bos, P.M.J., Gottardo, S., Scott-Fordsmand, J.J., Van Tongeren, M., Semenzin, E., Fernandes, T.F., Hristozov, D., Hund-Rinke, K., Hunt, N., Irfan, M.-A., Landsiedel, R., Peijnenburg, W.J.G.M., Sánchez Jiménez, A., Van Kesteren, P.C.E., Oomen, A.G., The MARINA risk assessment strategy: A flexible strategy for efficient information collection and risk assessment of nanomaterials. *Int. J. Environ. Res. Public Health*, *12*, 15007– 15021, 2015. https://doi.org/10.3390/ijerph121214961.
15. Nowack, B., Boldrin, A., Caballero, A., Hansen, S.F., Gottschalk, F., Heggelund, L., Hennig, M., Mackevica, A., Maes, H., Navratilova, J., Neubauer, N., Peters, R., Rose, J., Schäffer, A., Scifo, L., van Leeuwen, S., von der Kammer, F., Wohlleben, W., Wyrwoll, A., Hristozov, D., Meeting the needs for released nanomaterials required for further testing- the SUN approach. *Environ. Sci. Technol.*, 50, 6, 2747–2753, 2016. http://dx.doi.org/10.1021/acs.est.5b04472.

16. van Kesteren, P.C.E., Cubadda, F., Bouwmeester, H., van Eijkeren, J.C.H., Dekkers, S., de Jong, M.H., Oomen, A.G., Novel insights into the risk assessment of the nanomaterial synthetic amorphous silica, additive E551, in food. *Nanotoxicology*, 9:4, 442–452, 2015, doi: DOI: 10.3109/17435390.2014.940408.
17. Holden, P.A., Gardea-Torresdey, J.L., Klaessig, F., Turco, R.F., Mortimer, M., Hund- Rinke, K., Cohen Hubal, E.A., Avery, D., Barceló, D., Behra, R., Cohen, Y., Deydier- Stephan, L., Lee Ferguson, P., Fernandes, T.F., Herr Harthorn, B., Matthew Henderson, W., Hoke, R.A., Hristozov, D., Johnston, J.M., Kane, A.B., Kapustka, L., Keller, A.A., Lenihan, H.S., Lovell, W., Murphy, C.J., Nisbet, R.M., Petersen, E.J., Salinas, E.R., Scheringer, M., Sharma, M., Speed, D.E., Sultan, Y., Westenhoff, P., White, J.C., Wiesner, M.R., Wong, E.M., Baoshan, X., Steele Horan, M., Godwin, H.A., Nel, A.E., Considerations of environmentally relevant test conditions for improved evaluation of ecological hazards of engineered nanomaterials. *Environ. Sci. Technol.*, 50, 12, 6124–6145, 2016, http://dx.doi.org/10.1021/acs.est.6b00608.
18. Hund-Rinke, K., Herrchen, M., Schlich, K., Schwirn, K., Völker, D., Test strategy for assessing the risks of nanomaterials in the environment considering general regulatory procedures. *Environ. Sci. Eur.*, 27, 24, 2015, http://dx.doi.org/10.1186/ s12302-015-0053-6.
19. OECD, Guidance on Sample Preparation and Dosimetry: Series on the Safety of Manufactured Nanomaterials No. 36, 2012b, http://www.oecd.org/officialdocuments/publicdisplaydocumentpdf/?cote=ENV/JM/MONO(2012)40&docLanguage=En.
20. OECD, Preliminary Guidance Notes on Nanomaterials: Interspecies Variability Factors in Human Health Risk Assessment, Series on the Safety of Manufactured Nanomaterials No. 58, 2015, http://www.oecd.org/officialdocuments/publicdisplaydocumentpdf/?cote=env/jm/mono(2015)31&doclanguage=en.

2

Nanotechnology, Nanoparticles, and Living Systems

A ship is safe in harbor, but that's not what ships are for.

John A. Shedd

Abstract
Driven by demand for clean water, energy, healthcare, and advanced materials, nanotechnology is increasingly integrated into daily life. However, assessing its interaction with biological systems is complex. For example, nanoparticles used in drug delivery can cross the blood-brain barrier—offering benefits but also posing risks. Their lifecycle interactions with organisms present further exposure concerns. This chapter explores the need for applying regulatory standards to nano-based products and stresses the importance of ethical, environmental, and safety considerations in developing nanomedicine and related technologies.

***Keywords*:** Living-organisms, health-effect, inhaled, drug-carrier, harmful, environment, nano-adsorption, nano-filtration

2.1 Introduction

What nanotechnology is and the various applications it offers due to its unique physical and chemical properties have already been mentioned in Chapter 1. The specific and principal parameters of nanoparticles, such as their shape, size, surface characteristics, and inner structure, are bound to affect what happens when they come in contact with living systems. A living organism can encounter the nanoparticles as aerosols (solids or liquids in the air), suspensions (solids in liquids), emulsions (liquids in liquids), or even as fine powders. Moreover, the composition of specific nanoparticles used for different purposes can be very complex, depending on what interactions they have gone through with other chemicals or particles and

Madhuri Sharon. Risk, Safety, and Ethical Concerns in Nanotechnology, (19–42) © 2025 Scrivener Publishing LLC

on their lifetime. This is especially true when the chemical processes taking place on the surfaces of nanoparticles are also very complicated and are not completely known. They can interact differently, may remain free or group together depending on the attractive or repulsive interaction forces between them; for example, in a case where the nanoparticles are suspended in gas they tend to stick to each other more readily than in liquids. One of the concerns is not only the manufactured nanoparticles but also those that are formed through either the breaking down of larger particles or by controlled assembly processes. Many natural phenomena, human industrial, and domestic activities, such as cooking, manufacturing, road, and air transport, also release nanoparticles into the atmosphere. It is not clear how significantly human exposure to these nanoparticles has increased and is affecting us. One has to look into answering the following questions:

(i) *Can nanoparticles interact with living organisms?* One of the examples of the interaction of nanoparticles with the human system is when nanoparticles are administered dermally it was found that particle size was less important than the total charge in terms of permeation through the skin. For instance, only negatively charged particles were found to overcome the skin barrier and only when the concentration of charge was high enough.

(ii) *Which characteristics of nanoparticles are relevant for health effects?* The characteristics of nanoparticles that are relevant to health effects are obviously their very small size, chemical composition, and surface characteristics. The toxicity of nanoparticles depends on their chemical composition, and also the composition of any chemicals adsorbed onto their surfaces. However, the surfaces of nanoparticles can be modified to make them less harmful to health. Since the very small size of nanoparticles assists their crossing the blood-brain barrier, research is being done for their use in the delivery of drugs to the brain; however, one concern that remains post-delivery of drugs is the removal of unused accumulated nanoparticles in the brain. Lately, scientists have been trying to us biodegradable nanoparticles for drug delivery to the brain.

(iii) *How can inhaled nanoparticles affect health?* Materials that by themselves are not very harmful could be toxic if they are inhaled in the form of nanoparticles. One of the very dangerous sources of entry of nanoparticles into the human body is through inhalation. The source of such nanoparticles is polluted air. Air pollutants may not be nano size; we have to study whether the particles from the exhaust of vehicles is also as harmful as nanoparticles or more harmful than larger particles. The nanoparticles can be deposited throughout the human respiratory tract, including the lungs, and can potentially move from the lungs to other organs such as the brain, liver, spleen, and even a fetus. There is very limited data regarding these pathways, such as the actual number of particles that move from one organ to another and the impact of exposure time. Another possible route of entry of inhaled nanoparticles in the body is through the olfactory nerve ending in the nostrils. The nanoparticles may cross the mucous membrane inside the nose and then reach the brain or travel into the bloodstream through the olfactory nerve. The effects of inhaled nanoparticles in the body may include the lung inflammation and heart problems that occur due to their altering the system that regulates the involuntary functions in the cardiovascular system, such as control of heart rate.

(iv) *What are the health implications of nanoparticles when they are used as drug carriers?* Thousands of published studies are available on the use of nanoparticles as drug delivery vehicles in the human body which discuss how to get more of the drug to the target cells, how to reduce the harmful effects of the free drug on other organs, or how to circulate them long distances by evading the protection mechanisms of the body. Delivery is done either through injection, orally, or by applying onto the skin. There is the possibility of nanoparticles being recognized by the immune system and being eliminated. To avoid this the surfaces of nanoparticles are sometimes modified. Apart from drugs, nanoparticles are also being researched for delivery of genes to cells to treat cancer, as well as in vaccinations.

The harmful effect of nanoparticles along with drugs is due to their toxicity. Sometimes it is difficult to assess whether the toxicity is due to the nanoparticle or the drug; for instance, it has been shown that high concentrations of gold nanoparticles trapped in the liver can affect the function of this organ. Moreover, nanoparticles have the potential to cross the blood-brain barrier, which makes them extremely useful as a way to deliver drugs directly to the brain. On the other hand, this is also a major drawback because nanoparticles used to carry drugs may be toxic to the brain.

(v) *What are the effects of nanoparticles on the environment?* There are two prevailing views at this moment. One is that the use of nanotechnology will cause less chemical pollution, such as in auto-exhaust, etc., and the other is that nanoparticles will not only reduce environmental pollution but also create a new type of pollution.

(vi) *How should the harmful effects of nanoparticles be assessed?* All the traditional systems of assessing the harmful effects of chemicals may not be suitable for nanoparticles. For example, many traditional methods of measuring the harmful effects of chemicals are done by measuring the doses in terms of the mass because the harmful effects of any substance depend on the mass of the substance to which the individual is exposed. But for nanoparticles, it is more advisable to measure doses also in terms of the number of particles and their surface area because these parameters determine the interactions of nanoparticles with biological systems.

There are difficulties in assessing the harmful effect of nanoparticles; for instance, the use of nanoparticles as drug carriers may reduce the toxicity of the incorporated drug but it is sometimes difficult to distinguish the toxicity of the drug from that of the nanoparticle.

The various aspects of assessing the harmful effect of nanoparticles such as their characteristics, their distribution, effects on organs, immune and inflammatory systems are addressed in Chapter 4?

2.2 The Need for Nanotechnology in the Present-Day Scenario Cannot be Ignored

Almost all the branches of science have contributed to the advancement of nanotechnology and have benefitted in the form of useful applications (Figure 2.1). This is mainly because nanotechnologists believe that it offers stronger, lighter, cleaner, and "smarter" surfaces and systems which reduce environmental pollution, near-term expiration of Moore's Law of computing power (an impending crisis for Silicon Valley, anyway), and cures for diseases.

2.2.1 Increased Demand for Nanomaterial

The novel properties of nanomaterials such as high aspect ratio, high surface area reactivity, tunable pore volume, electrostatic, hydrophilic, and hydrophobic interactions, are useful in adsorption, catalysis, and filtration, thus providing high performance, affordable water, and wastewater treatment solutions both on a large as well as small scale.

Nano-adsorption is an effective measure of removing pollutants from water. Adsorption is a surface process wherein pollutants are adsorbed on a solid surface. Nano-adsorbents are used to remove inorganic and organic pollutants from water and wastewater. Parameters of a good adsorbent are small size, catalytic potential, high reactivity, large surface area, ease of separation, and a large number of active sites for interaction with different contaminants. Currently, carbon-based, metal-based, polymeric, magnetic, or non-magnetic oxide composite and zeolites are used for water treatment. Carbon-based materials are selective high-capacity sorbents for organic solutes.

Nanofiltration is a pressure-driven membrane separation process falling between ultra-filtration and reverse osmosis. Nanofiltration membrane is composed of carbon nanotube filters. Nanoceramic filters have been shown to remove turbidity, microorganisms, inorganic ions (Ca, Na), nitrates and heavy metals, such as arsenic, from groundwater and organic pollutants. Carbonaceous nanoparticles, such as CNT, CNF, graphene, etc., are suitable materials for constructing filtration systems.

Photocatalysis is another technique used in water purification. Photocatalysis is an advanced oxidation process that is based on the oxidative elimination of micropollutants and microbial pathogens. Most organic pollutants can be degraded by photocatalysis. Titanium dioxide (TiO_2) is the most popular and widely used photocatalyst. It exists naturally in three forms—rutile, anatase, and brookite—out of which anatase is used. When it is irradiated by UV light in the range of 200–390 nm, electron-hole (e^-/h^+) pairs are photoexcited. They move into the conduction and valent bands, which results in charge separation for an effective photocatalytic function depending on the redox potential of a substrate. TiO_2 coated onto vertically aligned MWCNT array on ceramic exhibits very low charge recombination of photoinduced electron-hole pairs and fast electron transfer from TiO_2 to MWCNTs, causing the effective killing of microbes [1]. Nanomaterials like dendrimers zeolites metal nanoparticles and carbonaceous nanomaterials are efficient nanocatalysts.

Many nanometals are effective biocide against microbes. Palladium-coated NPs is an effective catalyst for removing trichloroethane (TCE) from groundwater. Magnesium, gold, and silver nanoparticles act as antimicrobial, antibiotic, and antifungal agents. Zeolite NPs act as an effective sorbent (have successfully removed Cr(III), Zn(II), Ni(II), Cu(II), and Cd(II) from water) and ion exchange media for metal ions. They have a porous structure and can accommodate a variety of cations (Na^+, K^+, Ca^{2+}, Mg^{2+}).

2.2.2 Cheap and Clean Energy

Energy is one of the most pressing present and future global challenges. Nanotechnology is already being used to improve the efficiency of existing energy sources to produce advanced vehicles, scratch-proof eyeglasses, crack-resistant paints, anti-graffiti coatings for walls, transparent sunscreens, stain-repellent fabrics, self-cleaning windows, and ceramic coatings for solar cells. Research is being conducted on the use of nanotechnology to increase the safety of vehicles, to improve the adhesion of tires to the road, improve a car's body stiffness, apply ultrathin transparent coatings to displays or panes to avoid glare or condensation, etc. Products made with nanotechnology are being considered because they are faster, lighter, can access smaller spaces, are energy efficient, have unique properties at the nanoscale, and are cheaper.

2.2.2.1 Batteries

Batteries store electrical energy. In rechargeable batteries, the chemical process is reversible Batteries are used in transport, portable electronics, medical devices, power tools, and for storage of electricity produced by irregular renewable sources. Harvesting the potential of nanotechnology in renewable energy is approached through improving the components of batteries such as electrodes and electrolytes. Nanostructured materials increase surface area for electrolyte materials; nanoparticles enhance the conductivity and reduce the chance of a short circuit. Several types of nanomaterial allow for higher storage densities of lithium than standard metal or graphite electrodes such as carbon-coated silicon nanowires, carbon nanotubes, layered, nanostructured vanadium oxide, and manganese oxide. Coating the electrode's surface with nanoparticles, nanowires, or other nanostructures develops anodes with a greater density of locations to which lithium ions can attach, increasing the number of stored ions increases the stored electrical power. Thus, changing the atoms to which

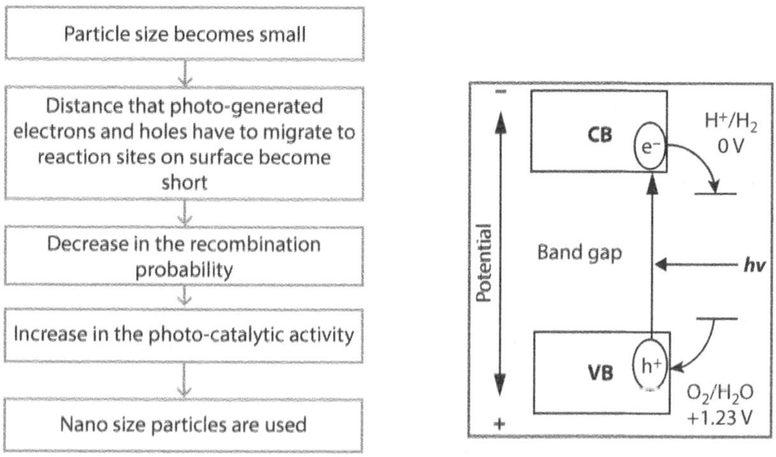

Figure 2.1 Photocatalytic water splitting using nano-sized particles.

the lithium bonds, thereby changing the electrochemical reaction, gives more energy, increasing the power. In electrolytes, nanoparticles added to solid polymer gel enhance the conductivity and storage capacity. Solid ceramics are used as they have high-temperature resistance and high-stress applications such as in large vehicles.

2.2.2.2 Fuel Cell

The fuel cell converts fuel directly into electricity in an electrochemical reaction. The limitations of fuel cells are that expensive materials such as platinum are needed for the electrode catalysts. Fuels other than hydrogen, which is costly and difficult to store, can cause fouling of the electrodes.

Therefore, platinum (Pt) nanoparticles are used instead of the solid platinum surface, thus allowing the use of much less Pt. To increase the accessibility of the Pt surface, PtNPs are supported on a porous surface.

2.2.2.3 Wind Energy

A wind energy system involves the conversion of kinetic energy into mechanical energy, which powers a generator and generates electricity. Nanocomposite using CNTs with excellent strength-to-weight and stiffness-to-weight ratios are used to construct longer more robust blades. Low-friction coatings and nano lubricants are used to reduce energy losses in gearboxes and thus further increase efficiency; and nano paints are used to increase the lifetime of wind turbines and reduce ice and dust buildup.

2.2.2.4 Hydrogen Energy

Hydrogen is a great source of non-polluting energy, making it an attractive transportation fuel. The power density of H_2 (156 atm) is 83.4 kWh/kg, which is much higher than propane (1.96 kWh/kg), methanol (1.45 kWh/kg), or ammonia (1.87 kWh/kg). Nanotechnology is being used for the safe and efficient storage of hydrogen. An attempt is being made to use carbon nanotubes (CNTs) and carbon nanofibers (CNFs) for the storage of hydrogen [2]. The generation of hydrogen is envisaged by photocatalytically splitting water, which is very effective and the cleanest way to do it. Figure 2.1 depicts photocatalytic water splitting using nanoparticle. Nanofluids based on molten salts are being explored for the storage of thermal energy. The main aim is to develop a nanofluid with a phase change behavior by adding different nanoparticles with suitable thermal conductivity and thermal capacity.

2.2.3 Food Industry

The food industry is another beneficiary of nanotechnology being used in processing and packaging. Food safety is improved by using packing material having antimicrobial-coated films and modifying the gas permeability required for different products to improve their shelf life.

2.2.3.1 Food Packaging Applications

Food packaging is one of the prevalent commercial applications of nanotechnology in the food sector. A polymer, such as polyolefins, polypropylene (PP), polyethylene (HDPE, LDPE), polyethylene terephthalate (PET), polystyrene (PS), or polyvinyl chloride (PVC), is used with nanomaterial as packaging material to improve physical performance, strength, stiffness, durability, and flexibility. In addition to providing resistance to attacks on food components, polymer-based nanocomposites also act as an oxygen and moisture barrier, and in the case of fresh fruits and vegetables, as a barrier to migration or gas diffusion. They can also provide the oxygen and carbon dioxide barriers in plastic containers of carbonated beverages, some of which are biodegradable. Polymer-based nanocomposites also offer better thermal properties, e.g., melting point, degradation, glass transition temperature, alteration in surface wettability, and hydrophobicity. Nanocoatings on the packaging surface (either the inside or the outside surface, or sandwiched as a layer in a laminate) are done to improve enhanced barrier performance. Nanocoating is done by vacuum-depositing aluminum on plastic films; organosilicate coatings on the surfaces of glass food and beverage containers (bottles, jars) using nanosilica provide high oxygen and moisture barrier films, thus preserving the aroma of the food and increasing the shelf life of packaged food. To control microbial contamination, antimicrobial nanomaterials (silver, zinc, titanium dioxide) and nanomaterial with biocide coatings are used. Attempts are being made to develop smart packaging by incorporating nanosensors to monitor and report on the condition of the food. These sensors can respond to environmental changes inside the package (temperature, humidity, level of oxygen exposure, degradation of product) and can detect certain chemical compounds, pathogens, and toxins in food, thus eliminating the need for inaccurate expiration dates. Biodegradable packing polymers are prepared using renewable biomass sources such as vegetable oil, corn-starch, potato-starch, or microbes, rather than fossil fuel plastics which are derived from petroleum. These polymers have increased gas and vapor barrier properties, better degradability, increased mechanical strength, and efficient antioxidants. Oxygen scavenging or antimicrobial bio-packaging increase food quality and safety. Barcode detection using nanoparticles is being developed for food safety.

2.2.3.2 Food Processing Applications

During food processing, nanoparticles have been applied to improve nutritional quality, flow properties, flavor, color, and stability or to increase shelf life. It is expected that in the future nanotechnology will help in developing healthier food with lower fat, sugar, and salts to overcome many food-related diseases. Effective olive oil hydrolysis using covalent immobilization of porcine triacylglycerol lipase onto functionalized nanoscale SiO_2 with reactive aldehyde group for better reuse, adaptation, and stability have been reported [3]. Several nano- and micro-structured assemblies of nanoparticles have been designed for encapsulation of food ingredients, additives, and nutrition. Nanocapsules containing desired flavors and colors that can burst at different microwave frequencies are being explored. Nanotubes made of milk protein by self-assembly are used as novel viscosifying and gelation agents. Nanoencapsulation is being used for the nutraceutical delivery of vitamins, antimicrobials, antioxidants, lycopene, and beta-carotene. Phytosterols and preservatives are available for enhanced taste, absorption, and bioavailability [4]. Whey protein-based nanospheres (40 nm), which are internalized by cells and degraded there to release the nutraceutical compounds, can be used as carriers for oral administration of nutraceutical agents to improve their bioavailability.

2.2.4 Defense Sector

The defense industry and research centers are spending millions of dollars to advance the use of nanotechnology. Some of the futuristic projects under development are given below.

2.2.4.1 Nano Armor

Armor fabric made of strong, lightweight, self–repairing nanomaterials for future combat suits and adaptive camouflage suits are being developed. For creating super-strong materials for protective gear, nano-tungsten is being used as the basic material. It is five times stronger than steel and at least twice as strong as any impact-resistant material, and can withstand the equivalent of dropping four diesel locomotives onto an area the size of a fingernail. Possible applications of nano-tungsten are ballistic protection personal body armor, bulletproof vests, flame and water-proof garments, vehicle armor, shields, helmets, and protective enclosures. Increased surveillance for better protection and safety is a major concern in defense. By using nano-fabric adaptive suits made of switchable fabric for improved thermal control, which is implanted with microsensors for body and brain sensing, environmental and situational awareness is possible [BBO NANO DEFENSE]. Material with ultra-hydrophobic surfaces (waterproof) is developed by depositing nanolayers of Teflon on Kevlar panels (material that is used for making bulletproof vests). In the offing is an ambitious project to make an invisibility cloak using optical negative-index metamaterials (NIM). Optical NIM can bend light in ways different from conventional materials. Beyond their invisibility ability, these structures also have applications in microscopes, circuits, and antennae.

2.2.4.2 Nano for Faster and Intensive Medical Help

Diagnostic support and therapy using nanotechnologies are being developed such as nano-fiber tourniquets, antitoxin protective equipment for use in biological warfare, lab-on-chip devices, *in-situ* tissue repair, and artificial organs. For soldiers in the field, it is critical to protect their health from regional diseases (e.g., malaria) and combat opportunistic local or systemic infections that are a constant threat. Hence, armor and fabrics used in uniforms are being made that can diagnose and treat a soldier in the field in the absence of doctors and hospitals. The need for loading smaller, faster nano-cameras and nano-detectors for online contacts is also considered very important, and can be an essential part of the future soldier's combat gear, including the smart helmet. This smart helmet consists of a helmet as a platform system equipped with an intelligent multi-sensor system for various tasks such as optical/IR camera (360° vision), radio frequency (RF) array antennas for positioning, radio frequency identification (RFID) and directed low power and efficient communication, acoustic arrays (microphones), B/C sensor arrays as an early warning system, and wireless EEG sensors. The use of nanotechnology for these sensors will reduce the weight of the equipment which is now attached or mounted on the helmet, creating less physical loading on the head. Anti-ballistic materials for the helmet are another essential need. Already, small-size, lightweight nanosensors with large reactive surface area are engineered, thereby providing an advantage over slower, more costly, laboratory-based analytical methods. Moreover, for the detection of multiple chemicals of chemical and biological warfare; chemical sensors that are capable of detecting a target chemical are chemically coated or decorated with functional groups that can recognize a specific chemical or biological agent.

2.2.4.3 Nanotechnology-Based Weapons

More effective, small, cheaper nanotechnology-based lethal as well as non-lethal weapons are needed with a focus on precision targeting, minimum weight and signature, optimal impact damage, and onboard intelligence. Non-lethal weapons are relatively new and evolving. They are made using energy waves in different forms directed by array technology. These materials with built-in condition and firing monitoring sensors are also super penetrator nanomaterials that sharpen upon impact to cause additional damage. They use high energetic propellants, e.g., nano dispersed aluminum as a propellant agent. For this purpose, quantum structures for small and efficient directed energy weapons, such as directed microwave and high-energy directed laser systems with anti-kickback systems, are being developed.

2.2.4.4 Lighter and Faster Aircraft

The present-day requirement for aircraft to be used in defense services is that they should be lighter, faster aircraft, with high strength nanocomposite plastics and biomimetic structures to reduce weight, and should have radar signature components with built-in condition and load monitoring sensors, such as fiber Bragg grating sensors, made with long-term, self-repairing, healing materials. Moreover, structures that adapt to changing conditions, such as adaptive aerodynamics and adaptive skin are also being envisaged along with stealth material design. Structures designed

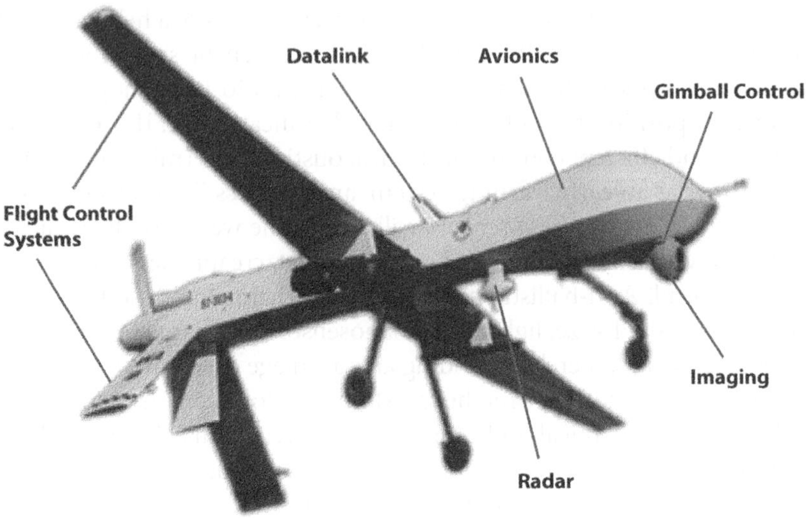

Figure 2.2 Unmanned aerial vehicle.

with radar absorption coatings, infrared thermal camouflage, and high energetic propellants, like nano-dispersed aluminum as a propellant agent, have already been developed. One of the achievements using nanotechnology in the aeronautic sector is the unmanned Nano Air Vehicle developed by AeroVironment, based out of Monrovia, CA, wha has developed the Nano Hummingbird under a DARPA research contract. It weighs no more than 10 g and can carry a payload of up to 2 g (Figure 2.2).

2.2.4.5 Nanotechnology for Vehicles

Apart from aerial vehicles, nanotechnology is used for motor vehicles as well as naval vessels. Nanocomposite plastics are replacing the metal components in lightweight and high strength vehicles. Smart components integrated into condition-based monitoring and load monitoring sensors, long-term self-repairing or self-healing materials, adaptive and camouflage active structures of flexible/rigid properties, nanofiber reinforced anti-ballistic structures, reactive nanoparticle armor, and shock-absorbing nanotubes are also being incorporated into vehicles. For naval vessels, adaptive aquadynamics are controlled using nanotechnology. Nanotechnology-based stealth and counter stealth technology for undetectable invisible warships and submarines are another landmark achievement of nanotechnology.

2.2.4.6 Radar-Absorbing Material (RAM)

To countermeasure against radar surveillance, RAM is employed in stealth for defense. The metallic surfaces of aircraft are coated with RAM to reduce the radar signatures of targets by the reduction of radar cross section. With the advent of nanotechnology, the RAM that is used now consists of a very thin absorbing layer (approximately a quarter wavelength at the frequency of operation, giving a 180° phase difference between the interface reflection and the emergent waves) of neoprene polymer sheets in which ferrite grains or CNTs are embedded.

2.2.4.7 Portable Energy/Power

With a focus on lightweight and energy-efficient power, the reduction of thermal, radar, and acoustic signature is an additional aspect for the military. The effort is ongoing to develop nanocomposite and high energetic material, lightweight efficient photovoltaic solar cell and μ-fuel cell and μ-thrusters to power portable electronic devices.

2.2.4.8 Nanotechnology for Satellites

Microsatellites are being used by the military for observation, inspection, anti-satellite weapons, and communication connected to information gathering and control systems, etc. The advantages of small, highly integrated modular satellites for military purposes are that they can be used as destruction satellites, spy satellites, and as a part of a swarm of satellites that are launched at the same time. Swarm formation flying can provide high-resolution observations. As compared to a minisatellite (50–500 kg) and microsatellite (10–50 kg) the nanosatellite is weighs only 1–10 kg. Nowadays, a picosatellite weighing <1 kg has also been developed. Nanotechnology provides lightweight, high strength nanocomposites having functionalities to incorporate nano-dispersed aluminum as propellant agent; nanomaterial enabled solar panels, deployers and actuators, μ-fuel cell, μ-thrusters for payload like radio communication, position, motion and guidance sensors, μ-instrumentation (x-ray, bolometer, optical and radio frequency (RF) interferometer, lab-on-chip, camera) and μ-radar, μ-bolometer (infrared) for security and guidance.

2.2.4.9 Nanotechnology for Nuclear, Chemical, and Biological Warfare

Warfare nuclear weapons and nano-built weapons are mostly the secret research and development efforts of the government concerned. Hence, not much information is available. Warfare has been an unavoidable aspect of human survival. The onset of nanotechnology in the defense sector, though resulting in better warfare techniques to reduce the mortality rates and the grievances associated with them, has ironically created more lethal weapons for their destruction. Nanotechnology has further solidified the need for safer yet more high-impact weaponry due to the unique properties associated with nanomaterials. In the right hands, this could enable a better tomorrow. In the wrong ones, there might not be a tomorrow.

2.2.5 Medical/Healthcare

It is interesting to note that it was a mathematician, Albert R Hibbs, who pointed out the idea of nanomedicine to Feynmann (the father of nanotechnology), saying that "It would be interesting in surgery if you could swallow the surgeon. You put the mechanical surgeon inside the blood vessel and it goes into the heart and "looks" around.... It finds out which valve is the faulty one and takes a little knife and slices it out. Other small machines might be permanently incorporated in the body to assist some inadequately-functioning organ." That possibility is becoming a reality now. Nanotechnology has entered the arena of Biology and Medicine in a wide variety of ways:

- Drug delivery and development;
- Therapeutics, including gene therapy and scaffolds for tissue engineering;
- Diagnosis of diseases and abnormal conditions using DNA probes and nanoscale biochips;
- Surgery;
- Robotics.

Moreover, applications in healthcare have offered many other advantages such as increasing knowledge about the human body.

2.2.5.1 Application in Drug Delivery

Nanotechnology helps in targeted drug delivery, thereby reducing the dose and limiting the side effects of therapeutics while making them more effective. It also helps in making gene, protein, and peptide delivery easier. A plethora of applications of nanoparticles for cancer therapy are very effective; for example, nanometals like gold nanoshells could fight cancer because of their ability to absorb radiation at certain wavelengths. Once a target-delivered nanoshell enters tumor cells and radiation treatment is applied, they absorb the energy and heat up enough to kill the cancer cells. Besides gold, many other nano elements can also be used.

2.2.5.2 Nanodiagnostics Application

Nanotechnology has offered faster diagnosis using a minimal amount of tissues or metabolites, and even artificial muscles, nerves, and bones. Medical diagnosis covers all fields of science for imaging, measuring, and manipulating matter at the nanoscale that acts as a support system to prevent and treat life-threatening diseases. The suitability of nanomaterial in diagnostics is due to its properties like higher chemical reactivity, increased mechanical strength, and faster electrical and magnetic responses, owing to its high surface to unit volume ratio and easy functionalization for attachment to biomolecules, allowing detection of disease biomarkers in a lab sample at a very early stage. Because of their small size, nanomaterials can readily interact with biomolecules and gain access to many areas of the human body by passing through intracellular spaces.

2.2.5.3 Application in Surgery

With the help of nanotechnology, minute surgical instruments and robots can be made which can be used to perform microsurgeries on any part of the body. Instead of damaging a large amount of the body, these instruments can be precise and accurate, targeting only the area where surgery should be done. Visualization of surgery can also be improved. Instead of a surgeon holding the instrument, computers can be used to control the nano-sized surgical instruments with the support of nano cameras that can provide close-up visualization of the surgery. This lessens the chance of any errors occurring. Moreover, surgery could also be done on tissue, genetic, and cellular levels.

2.2.5.4 Application in Medical Robotics

With nanotechnology, minute surgical instruments and robots can be made which can be used to perform microsurgeries on any part of the body. Instead of damaging a large amount of the body, these instruments would be precise and accurate, targeting only the area where the surgery should be performed. Visualization of surgery can also be improved. Instead of a surgeon holding the instrument, computers can be used to control the nano-sized surgical instruments along with the support of nano cameras that can provide close-up visualization of the surgery. This will decrease the chance of any errors occurring. Moreover, surgery could also be done at tissue, genetic, and cellular levels.

Potential applications of nano-robotics, in medical fields include early diagnosis and targeted drug delivery for cancer, biomedical instrumentation, surgery, pharmacokinetics, monitoring of diabetes, and healthcare. It is envisaged that in the future medical nanotechnology will employ the injection of nanorobots into the patient to perform treatment at a cellular level.

Apart from these services in human healthcare, nanotechnology also provides snapshots of the human body for a better understanding of how it works; and the workings of cells, bacteria, viruses, etc., can also be better explored. The causes of relatively new diseases can be found and prevented. Even genome sequencing can be made much easier. Biological causes of mental diseases can be monitored and identified. Tissue engineering could also be done using nanomaterials, especially to make use of artificially stimulated cell proliferation by using suitable nanomaterial-based scaffolds and growth factors. Advances in nanotechnology-based tissue engineering could also lead to life extension in humans and other animals.

2.2.6 Nanotechnology in Agriculture

Needless to say, agriculture is the backbone of human survival. With the increasing population, the challenge faced by Agrotechnology is to provide more, safe, healthy food. Disease prevention is another threat to agriculture. Nano-agrotechnology is another arena where many nano-based products on the market use nanomaterials like CNTs, nano-silver, nano-silica, nano-alumina silicate, etc. Intensive research is going on in the following areas.

2.2.6.1 Precision Farming

Improved precision farming:

- (i) Maximizes crop yield. Various non-traditional methods for the slow release of NPK are now on the market;
- (ii) Reduces production cost by minimizing the input, i.e., fertilizers, using efficient formulations with a reduction in fertilizer dose;

(iii) Reduces spray drift with aerodynamic nanoparticles for optimized deposition during foliar spray, especially of hormones and trace elements. Alumino-silicate nanotubes are used as a foliar spray, as they are easily picked up in insect hairs on the plant surface. Insects actively groom and consume pesticide-filled nanotubes. They are biologically more active and relatively more environmentally safe pesticides. Mesoporous spherical silica NP can deliver DNA and chemicals into plants, thus creating a powerful new tool for targeted delivery into plant cells. Porous Si-NP spherical powders are used to introduce DNA and chemicals into *Arabidopsis*, tobacco, and corn plants.

(iv) Uses controlled release advanced delivery systems for season-long treatment with herbicides, pesticides, fungicides, and rodenticides to reduce the chance of pathogen attack. Nanosilver is known to have strong inhibitory and bactericidal effects and a broad spectrum of antimicrobial activities. Its high surface area and high fraction of surface atoms, imparts a high antimicrobial effect, and more adhesive on bacteria and fungus. Colloidal nanosilver (1.5 nm average diameter) solution is used against rose powdery mildew caused by *Sphaerotheca pannosa*. It also eliminates unwanted microorganisms in planter soil and hydroponic systems. It is being used as a foliar spray to stop fungi, mold, rot, and several other plant diseases. Aqueous silicate solution is known to be absorbed into plants to increase disease resistance and stress resistance. It promotes the physiological activity and growth of plants and induces disease and stress resistance in plants. Porous hollow silica nanoparticles (PHSNs) are loaded with validamycin. Nano-Si is also used in combination with nano-Ag to control various plant diseases, as it is an antifungal and controls powdery mildews of pumpkin. Nano-Si/Ag composite has been found to inhibit 100% growth of *Pythium ultimum, Magnaporthe grisea, Colletotrichum gloeosporioides, Botrytis cinerea;* and *Rhizoctonia solani, Bacillus subtilis, Azotobacter chroococcum, Rhizobium tropici, Pseudomonas syringae,* and *Xanthomonas compestris pv. vesicatoria.* However, at higher concentrations, they cause chemical injuries. Other biocidal nanoparticles used in agriculture are aluminum oxide (ANP), zinc oxide, and titanium dioxide (to control rice weevils, etc.).

Copper nanoparticles in soda-lime glass powder are another efficient antimicrobial and antifungal material.

(v) Treats seeds to improve plant health and stress resistance. Very strong CNTs are used as they not only penetrate the hard coat of germinating seeds, but also exert a growth-enhancing effect when used as a vehicle to deliver desired molecules into the seeds during germination that can protect them from diseases without having a toxic or inhibitory effect on the plant.

(vi) Has popularized the use of nanoparticle-coated greenhouse covering materials for photoprotection.

(vii) Reduces run-off through monitoring environmental variables and applying targeted action. Nanosensors can detect herbicides by checking the oxygen level in the soil. Nanoparticles are used to attack the seed coating of weeds to stop germination.

Looking over the list of nanotechnology's applications in agriculture, it is not possible to ban the growth of this field. It has both economic as well as commercial potential. Therefore, it is essential to assess the potential hazards associated with this technology and its corresponding products in emerging areas for their health risks. At the initial stages of these healthcare applications, many side effects and the darker side of nanotechnology have not been seriously considered. However, this topic of concern is being globally studied to minimize the risks. During the last decade, several research and production strategies and challenges have been proposed for safe handling to avoid exposure to nanoparticles and the effects of nanotechnology in general. But the major problem is that all the different nanoparticle types cannot be effectively evaluated for their safety and environmental effects and will require lots of effort and time. This is especially true because they affect all living organisms—humans, animals, plants (both aquatic and terrestrial), and the food cycle—thus, the environment as a whole.

As this cross-disciplinary technology matures and flourishes, concerns regarding the potential health effects of exposure to untested materials increases; because many newly developed products have not been adequately evaluated. Therefore, it is important to ensure that societal and commercial advantages are not outweighed by potential human health or environmental disadvantages.

We also come across articles like "Nanotechnology: From "Wow" to "Yuck"?" by Kristen Kulinowski [5] from the same Rice University where one of the great advocates of nanotechnology, Professor Richard E. Smalley, came.

Like GMOs, this field is also facing critical comments from those who want to accept nanotechnology with a pinch of salt. There is growing public concern over nanotechnology's benefits and risks, though we have to accept that it is no longer a futuristic science. Nanotechnology-based products showing great promise are already on the market; and this has generated controversy and societal as well as ethical concerns which have been overlooked in the enthusiasm for developing these products. This demands the need for a debate on ethical concerns, as was done for GM food, which clarified many issues. These debates will be more complicated from an ethical point of view because nanotechnology encompasses many fields of science, technology, and engineering.

Let us now take a brief look at the interdependence and interactions between living organisms, both plants and animals, and the possibility of how one being affected by nanoparticles, may become a concern for the other.

2.3 Interdependence and Interaction between Various Living Systems

There may be studies suggesting that the use of nanoparticles should be stopped for human consumption but can be used in abiotic activities or for agriculture. But the problem is that there is interdependence and interaction between various living as well as non-living organisms. A nano-paint or non-fertilizer may at some point or the other interact with humans, creating exposure challenges concerning nanoparticles.

Living organisms encompass humans, animals, plants, microbes, and viruses. They all are in some way or other related to each other. Human beings especially depend on plants and animals for everyday needs such as food, shelter, medicine, clothes, and even microbes. Some interactions with microbes and viruses may not be very fruitful and rather damaging, but we cannot deny their presence and interaction with our lives. Therefore, if any of these systems are affected by nanoparticles or nanotechnology, they will be a big risk to mankind. Our food chain or food cycle is one of the greatest factors that correlate with us.

2.3.1 Life-Cycle and Exposure Challenges

Chances are that under specific conditions more hazardous nanomaterials may be formed during the life cycle due, for example, to disintegration of a coating or photoactivation, which may lead to more hazardous nanomaterials than their pristine counterpart. Changes in nanomaterials during their life cycle need risk assessment by measuring the nanomaterials and their physicochemical properties such as degree of agglomeration and size distribution in complex matrices. The development of computational models to estimate changes in physicochemical properties during the life cycle caused due to exposure is therefore highly recommended.

Bioaccumulation or buildup of nanoparticles poses an ecotoxicity risk to human health [6–10]. It is particularly important for those nanomaterials that are very slow in being eliminated from tissues. It should be noted that bioaccumulation of hydrophobic nano-material is different from the bioaccumulation of hydrophobic non-nanosubstance. The behavior of nanomaterials cannot be described by chemical equilibrium. As the octanol-water partitioning coefficient assumes an equilibrium between the dissolved concentrations in the octanol and water phases, it is not a predictor for bioaccumulation or the environmental fate of nanomaterials [11], or even for human health consideration, where key points/accumulation may be different in human tissues affecting and linking toxicokinetic data to physicochemical properties of the nanomaterial and their dissolution rate in physiologically relevant media. Moreover, the long-term effects of bioaccumulation *in-vitro* and *in-vivo* deserve special attention, as they may vary.

There is very little data on exposure route/media extrapolation for nanomaterials [12]. The route/media for exposure, which is envisaged to be *via* lung (air), gastrointestinal tract (gastrointestinal fluids), skin, and intravenous applications, differs greatly based on exposure (stable dispersions, relevance of high dose, internalized dose).

2.4 Concerns of Nanoparticle Properties Impacting Interaction with Living System

Interaction between nanoparticles and living systems is affected in general by the surface effect, size and shape effect, and solubility persistence. However, the physicochemical properties of nanomaterials that are relevant for their interaction with a living system result in potential risk and may be altered during their life cycle [12]. Specific studies have only appeared recently and are still scarce. Most of the information dealing with the toxicity of nanoparticles is from studies on inhaled nanoparticles and from pharmaceutical studies in which nanomaterials are used in drug delivery.

2.5 Concluding Remarks

Generous funding by many countries has been looking at the pros and cons of nanotechnology. I cannot say all the countries that are investing in nanotechnology have created laws for its use, disposal, etc., though some countries have aimed their efforts in this direction. It is being realized that similar regulations as for chemical or medicinal uses can be applied to nano-based products. Nanotechnology-based healthcare needs very thorough ethical, environmental, and safety considerations. Scientists are currently researching better electronics and safe materials for nanomedicine alongside nanomedicine itself. The technology is being regulated by the government with the primary concern being the safety of the people. We know that nanomedical technology, if used carefully and meticulously, will guarantee great benefits. But one of the issues of nanotechnology in medicine will be how many nanotechnology implementations in the human body (to protect against diseases and offer "enhanced immunization") would it take for a human to no longer remain human? There is a fear of a decrease in the gap between humans and robots in the distant future. Another aspect that we cannot ignore are the economic issues regarding nanotechnology in medicine. If it is very expensive it may widen the gap between the rich and the poor. Bioethical researchers believe that nanomedicine could be manipulated to harm the human body rather than heal it. What if nanomedicine technology is used for terrorism purposes? There certainly are more questions than adequate answers!

For protective measures against the ethical and risk concerns from nanoparticles and nanotechnology; nanotechnologists have to address questions regarding (i) the entry routes into the human body (whether indirect entry through lungs by inhalation or skin, or direct entry with nanotheranostics); (ii) anatomical aspects caused by penetration at the cellular level; (iii) biochemical aspects like interaction with metabolism; (iv) cytogenetical aspects due to interaction with DNA; and (v) toxicological effects. In the forthcoming chapters of this book, we will be discussing these concerns and will also be looking at whether the current risk assessment methodologies for nanoparticles are adequate; as well as whether existing methodologies to assess the potential human health and environmental risks associated with products of nanotechnology are appropriate.

Nanotechnology, I love thee with all thy faults.

References

1. Oza, G., Pandey, S., Gupta, A., Shinde, S., Mewada, A., Jagadale, P., Sharon, M., Sharon, M., Photocatalysis-assisted water filtration using TiO_2-coated super-aligned multi-walled carbon nanotube array for the removal of *Escherichia coli 0157:H7.*
2. Sharon, M., Sunil, B., Sandesh, J., Sathiyamoorthy, D., Dasgupta, K., Sharon, M., hydrogen adsorption by carbon nanomaterials from natural source. *Asian J. Exp. Sci.*, 22, 2, 75–88, 2008.
3. Bai, Y.X., Li, Y.-F., Yang, Y., Yi, L.-X., Covalent immobilization of triacylglycerol lipase onto functionalized novel mesoporous silica supports. *J. Biotechnol.*, 125, 4, 574–582, 2006.
4. Momin, J.K., Jayakumar, C., Prajapati, J.B., Nutrition And Food Science Potential of nanotechnology in functional foods. *Emir. J. Food Agric.*, 25, 1, 10–19, 2013, doi: DOI: 10.9755/ejfa.v25i1.9368 http://www.ejfa.info/.
5. Kulinowski, K., Nanotechnology: From "Wow" to "Yuck"? Bulletin of Science Technology & Society, February 2004, DOI: 10.1177/0270467604263112.
6. Bos, P.M., Gottardo, S., Scott-Fordsmand, J.J., van Tongeren, M., Semenzin, E., Fernandes, T.F., Hristozov, D., Hund-Rinke, K., Hunt, N., Irfan, M.A., Landsiedel, R., Peijnenburg, W.J., Sanchez Jimenez, A., van Kesteren, P.C., Oomen, A.G., The MARINA risk assessment strategy: a flexible strategy for efficient information collection and risk assessment of nanomaterials. *Int. J. Environ. Res. Public Health*, 12, 15007–15021, 2015.
7. Dekkers, S., Oomen, A.G., Bleeker, E.A., Vandebriel, R.J., Micheletti, C., Cabellos, J., Janer, G., Fuentes, N., Vázquez-Campos, S., Borges, T., João Silva, M., Prina-Mello, A., Movia, D., Nesslany, F., Ribeiro, A.R., Emílio Leite, P., Groenewold, M., Casse, F.R., Sips, A.J.A.M., Dijkzeul, A., van Teunenbroek, T., Wijnhoven, S.W., Towards a nanospecific approach for risk assessment. *Regul. Toxicol. Pharmacol.*, 80, 46–59, 2016. http://dx.doi.org/10.1016/j.yrtph.2016.05.037.
8. Arts, J.H., Hadi, M., Irfan, M.A., Keene, A.M., Kreiling, R., Lyon, D., Maier, M., Michel, K., Petry, T., Sauer, U.G., Warheit, D., Wiench, K., Wohlleben, W., Landsiedel, R., A decision-making framework for the grouping and testing of nanomaterials (DF4nanoGrouping). *Regul. Toxicol. Pharmacol.*, 71, S1–27, 2015.
9. Arts, J.H., Irfan, M.A., Keene, A.M., Kreiling, R., Lyon, D., Maier, M., Michel, K., Neubauer, N., Petry, T., Sauer, U.G., Warheit, D., Wiench, K., Wohlleben, W., Landsiedel, R., Case studies putting the decision-making framework for the grouping and testing of nanomaterials (DF4nanoGrouping) into practice. *Regul. Toxicol. Pharmacol.*, 76, 234–261, 2016.
10. Hansen, F.S., Alstrup Jensen, K., Baun, A., NanoRiskCat: a conceptual tool for categorization and communication of exposure potentials and hazards

of nanomaterials in consumer products. *J. Nanopart. Res.*, 16, 2195, 2014. http://dx.doi.org/10.1007/ s11051-013-2195-z.

11. Oomen Agnes, G., Steinhäuser, K.G., Bleeker, E.AJ., van Broekhuizena, F., Sipsa, A., Dekkersa, S., Wijnhovena, S.WP., Sayrec, P.G., Risk assessment frameworks for nanomaterials: Scope, link to regulations, applicability, and outline for future directions in view of needed increase in efficiency. *NanoImpact*, 9, 1–13, 2018.
12. Stefania, G., Mech, A., Drbohlavová, J., Małyska, A., Bøwadt, S., Sintes, J.R., Rauscher, H., Towards safe and sustainable innovation in nanotechnology: State-of-play for smart nanomaterials. *NanoImpact.*, 21, 100299, 2021.

3

Methodologies for Assessing Risks Associated with Human Health and the Environment by Nanotechnology-Based Products

Ethics and science need to shake hands.

Richard Clarke Cabot

Abstract

This chapter highlights the need for reliable methods to assess the health and environmental risks of nanotechnology. It focuses on exposure to engineered nanoparticles in workplaces, the environment, and across biological systems. Traditional chemical risk assessment methods, such as those for PM 2.5, offer a baseline, but new tools—like Control Banding and Risk Prioritization—are recommended. The chapter also addresses limitations of conventional industrial hygiene instruments and stresses the importance of accurate measurement and analysis of airborne, gaseous, and liquid nanoparticle exposure.

Keywords: Human-health, workplace, inhalation, dermal-exposure, skin, risk-assessment, particulate-matter, oral-ingestion

3.1 Introduction

With the continuous increase in many engineered nanomaterials as well as nanoparticle-associated production volume, an assessment of the concerns regarding their potential risk to the environment and human health is necessary. The question of how the harmful effects of nanoparticles should be assessed raises obvious concerns about whether traditional methods for protecting the environment and human health from the risks caused by the physical and chemical impacts of engineered materials will be useful for all

Madhuri Sharon. Risk, Safety, and Ethical Concerns in Nanotechnology, (43–62) © 2025 Scrivener Publishing LLC

nanomaterial or not. Traditionally, for chemicals, dose measurement is done in terms of mass to which the individual is exposed. But for nanoparticles, it may not be reasonable, because due to their enhanced surface-to-volume ratio the interaction of nanoparticles with the biological system gets altered. To create a standard methodology for all nanoparticles poses lots of uncertainties due to their varying sizes, shapes, internal structures, and applications. Therefore, the harmful effects of nanoparticles to which the individual as well as environment are exposed, should be measured in terms of several particles and their surface area because these parameters further determine the interactions of nanoparticles with living systems. In human beings the characteristics of nanoparticles, their distribution, and their effects on organ systems, including effects on the immune system, will be varied. Moreover, conclusions based on tests on healthy animal models may not be suitable for human beings as some of the effects of nanoparticles may only be a risk for susceptible organisms and predisposed individuals, but not to healthy people.

Prior to assessing the risk of human beings being exposed to nanoparticles, it is important to know how they may affect the environment. Comparatively less research has been done on the effects of engineered nanoparticles on animals and plants in the environment. Research has been focused more on examining the uptake and impact of nanoparticles at a cellular level in humans [1], animals [2] and microbes [3]. There is a need for. examining and interpreting the existing data, particularly the persistent insoluble nanoparticles that may cause much greater problems in the environment than those revealed by human health assessments. These risk assessments and management have to be done at the workplace where research and production are being done, as well as in the environment (including plants, animal, and humans) where they are used and disposed of.

3.2 Workplace Exposure to Effects of Engineered Nanoparticles

Determining the effects of engineered nanoparticles on those in the workplace encompasses the interface between the environment and the individual or group. It involves collecting and systematically organizing information on the nanoparticles and other hazardous materials used (ergonomic hazards, chemical agents, and physical agents) as well as the workforce at the workplace (ventilation characteristics, use of personal protective equipment and procedures). Classification of workers is based on distribution of exposure and then measured by using appropriate instruments and models. The understanding of the workplace, workforce, and agents in the work environment, and potential of exposures helps in properly deciding and executing the strategies required to develop methodologies. In a work place the exposure routes to the workforce is through

(i) Inhalation that affects the respiratory tract, i.e., tracheobronchial and alveolar regions of the lungs. We know that a variety of nanoparticles that are found in indoor and outdoor settings causes respiratory diseases [4]. Here, particle size and duration of exposure is a big consideration. Hence, occupational exposure limits (OEL) need to be checked for different nanoparticles. However, it must be stated here that the majority of chemicals in commercial use do not have an established OEL or permissible exposure limits (PELs). Repeated daily workplace exposure to a chemical by inhaling it at an airborne concentration below its OEL is unlikely to lead to harmful health effects in most workers. However, many thousands of chemicals are in use, and there is not an OEL for every chemical, chemical mixture, fume, or emission.

(ii) Oral ingestion of nanoparticles is a rather easy route of uptake of nanoparticles because they are efficiently absorbed through the gastrointestinal tract [5]. There is a permissible limit particle, however, for dust permissible nanoparticle needs more thorough studies. Different countries are working on deciding the benchmark exposure limits for nanoparticles. For example, the German Institute for Occupational Safety and Health (the IFA) and the British Standards Institute (BSI) in the United Kingdom for various nanoparticles, such as CNT, fullerene, metal oxides (TiO_2, ZnO, cerium oxide, iron oxide, antimony oxide, tin oxide), silica, gold, iron, silver, cobalt, beryllium, arsenic, lead, nickel, cadmium-containing quantum dots, zinc chromate, chromium, dendrimers, polystyrene, nanoclay, etc.

(iii) Dermal exposure is also considered to be a cause of concern, especially through cuts, wounds, and lacerations, which can facilitate dermal penetration [6]. The composition of nanoparticles, surface coating, and geometry influences the penetration through the skin. Moreover, formulations of skin creams that are supposed to be taken up by the dermal layer by altering and enhancing the skin's permeability and its barrier properties; such as dimethyl sulfoxide, may facilitate the absorption of nanoparticles by removing much of the lipid matrix of the stratum corneum (Figure 3.1).

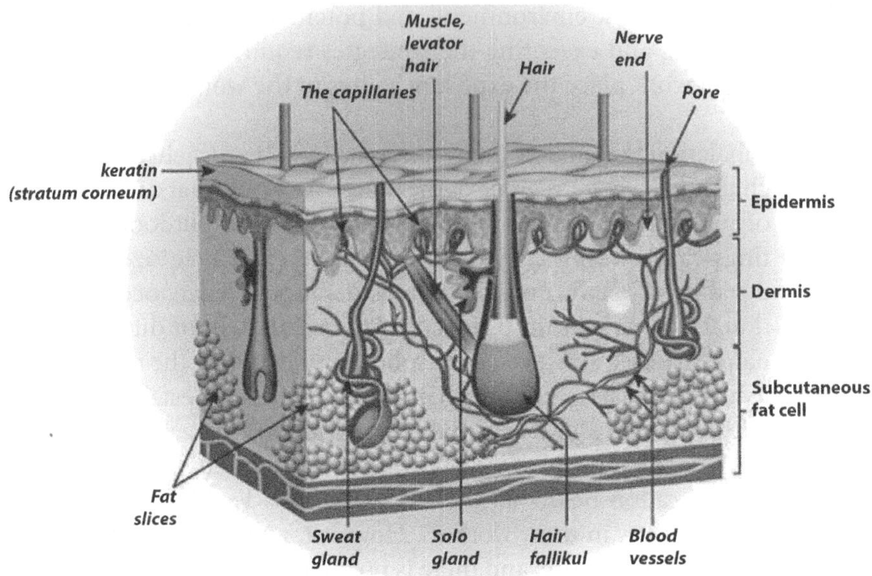

Figure 3.1 Schematic three-dimensional view of human skin anatomy.

3.3 Effects of Engineered Nanoparticles on Animals, Plants and Environment

Since 2008, ENRHES (Engineered nanoparticles: Review of health and environmental safety) projects and many research groups have been actively assessing, collecting data on the potential impact of engineered nanoparticles on health, safety, and the environment, especially fullerenes, carbon nanotubes, metals and metal oxides. There have been many reports assessing the impact of nanomaterial, which, in a nutshell, indicate that every newly engineered nanoparticle needs a thorough assessment of its impact. Aschberger *et al.* [7] studied the impact of carbon- and metal oxide-based nanomaterials on aquatic plants (algae) and aquatic water flea *Daphnia*. They concluded that the main risk to the environment is expected from metals and metal oxide, especially for algae and *Daphnia*. The main risk to human health may arise from chronic occupational inhalation exposure, especially during activities of high particles release and uncontrolled exposure. It is important to consider that each type of metal to be considered will have different forms, shape and size; for example, nanosilver may be in crystalline form, functionalized particle, and pure silver, and these different forms can exhibit quite different properties and impacts.

3.4 Traditional Methods of Risk Assessment of Chemicals

Usually, many follow the general approach specified in the REACH Guidance on Information Requirements and Chemicals Safety Assessment [8], which is similar to the format for preparing a Chemical Safety Report under REACH. This involves following standard risk assessment: (i) hazard identification, (ii) hazard characterization, (iii) exposure assessment, and (iv) risk characterization. However, it must be said that it has not yet been applied to nanoparticles. At the moment, there are no official guidelines on what is an appropriate testing procedure.

The conventional and standard methods of risk assessment caused by chemicals (PM 2.5) has been a very complex situation that had a lack of sufficient information about multiple components. The risk framework has evolved over a decade with the help of various agencies that resorted to the use of default assumptions and extrapolations to fill in the data gaps. The traditional methods of risk assessment have been the basis for the risk assessment of nanoparticles. In the case of nanoparticle health risks, the uncertainties at the moment are still very large. To understand and gain insight into the risk assessment of nanoparticles it is important the reader should first have an idea about the traditional methods of risk assessment that are employed for pharmaceutical products and then evaluate whether it can be applied to nanomaterials. An assessment of the hazards of nanoparticles has to be based on the level of particulate matter (PM 2.5).

3.4.1 What is PM 2.5?

Here, PM represents particulate matter and PM 2.5 is the fine particulate matter of <2.5 μm that is considered as a pollutant. If it is present in the air at high levels then it becomes a cause for concern for people's health. The presence of a high level of PM 2.5 reduces the visibility. It is a complex mixture of solid and aerosols of small droplets of liquid and solids and a solid core with liquid coatings. According to the Giannadaki *et al.* [9], implementing the US air quality standard for PM 2.5 worldwide can prevent millions of premature deaths per year. Their studies have reflected the need to adopt stricter limits for annual mean PM 2.5 levels globally to 12 $\mu g/m^3$ or an even lower limit to substantially reduce premature mortality in most of the world. But if the level exists for 24 hours at >35 $\mu g/m^3$ then the air is accepted as unhealthy for causing respiratory problems like asthma. Use of a PM 2.5 filter inside a regular mask is recommended for breathing protection.

Since the COVID-19 pandemic, the use of N95 respirators with PM 2.5 filters are being recommended, because N95 masks can filter 95% of all 0.3-micron particulates.

A recent work by Zhang *et al.* [18] directed towards developing an inexpensive simple method to remove indoor respiratory aerosols is based on the fact that over 95% of respiratory aerosols are electrically charged. To efficiently remove these charged aerosols, they have designed a high-voltage supply (that is widely used in electric fly zappers) and a parallel-plate to capture respiratory aerosol capturer. It has a sampling mask for respiratory aerosols. The mask is a modified 3M mask (Model 3N11CN), with the exhalation valve filter removed and connected to subsequent sampling instruments using a 3/8-inch stainless steel tube.

3.4.2 Parameters and Concerns for Formulating Methods for Risk Assessment

Parameters are the major basis for quantitative risk assessment values. After considering the parametric uncertainty associated with observed value models, many important relationships between variables and their functional forms are also faced. The environmental health framework for

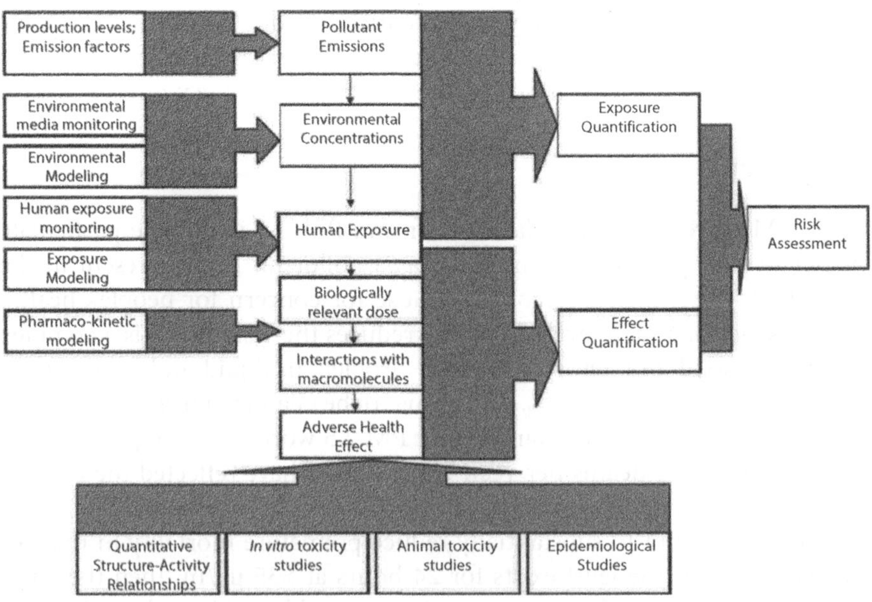

Figure 3.2 General environmental health framework for risk assessment. (Source: Journal of Nanoparticle Research, 2007, 9: 137-156. Copyright © Springer).

risk assessment involves (i) the relationship between quantitative structure and activity, (ii) *in-vitro* toxicity assessment, (iii) *in-vivo* animal toxicity studies, and (iv) epidemiological studies.

More than a decade ago, Kandlikar *et al.* [10] have nicely presented the concerns and framework of risk assessment (Figure 3.2).

3.4.3 Challenges Related to the Traditional Industrial Hygiene Approach

It can be said that many uncertainties will be faced if traditional exposure assessment strategy is applied to nanotechnology-based products, processes, and nanomaterials; because (a) so far nanotechnology processes are not well documented in many workplaces, and (b) the same properties of nanoparticles that are desirable for a particular application are often biologically active and toxic (e.g., depending on the size <10 nm) and may obey quantum physics laws rather than classic physics and exhibit physicochemically unique optical, magnetic, and electrical characteristics because as the diameters of particles decrease the proportion of atoms or molecules on the surface increases. Due to these concerns the toxic effects of nanomaterials on humans remain poorly understood.

3.4.4 Traditional Instruments for Assessing Exposure

In traditional methods, personal exposure to airborne particles within the breathing zone is sucked through a filter mounted inhalable commercially available portable sampler. From the collected sample the mass concentration is then computed and determined gravimetrically or by chemical analysis. Various direct-reading instruments are used for measuring particle concentrations; such as:

- *CPC (condensation particle counter):* Measures *total number concentration* from ~15 nm to ~1 ECHA (https://echa.europa.eu › documents › mb_17_2...Doc.: MB/17/2009 final. General Report of the. European Chemicals Agency. 2008. 24 April 2009. (Art. 78(a), 83(3) of Regulation (EC) No. 1907/2006).
- *OPC (optical particle counter):* Measures *number concentration* by size from ~300 nm to ~10 μm.
- *Photometer:* Measures *mass concentration* from ~300 nm to ~10 μm. The direct readout of mass concentration can be

operated with a size separator on the inlet to provide respirable mass concentration, PM_{10}, PM 2.5 or other size fractions.
- *Diffusion Chargers:* Generally record *surface area concentration* of sub-μm particles. The diffusion charger operates by producing positive ions with an electrical corona attached to the surface of particles, and the charged particles are collected on a grounded filter. The electrical current is drained from the filter and measured with a highly sensitive electrometer, which is related to the particle surface area concentration.

All these instruments are available as handheld instrument models. Moreover, some time-integrated measurements are done for detailed characterization. For example:

- TEM (transmission electron microscopy) and SEM (scanning electron microscopy) are used to analyze size and morphology;
- TEM with EDS (energy dispersive spectrometry) is used for assessing the chemical composition of particles [11];
- ICP-MS (inductively coupled plasma mass-spectrometry) is used for routine monitoring of bulk chemical analysis of the collected nanoparticles.

3.5 Methods for Risk Assessment of Nanoparticle

The exposure assessment strategy in nanotechnology demands basic characterization. Such analysis of workplace and workforce should include gathering information on

(i) *Processes and Process Flow:* Accounts for raw material storage, material transfer during dumping, conveying, and bagging. Post-production process output, i.e., intermediate, final, and byproducts, along with cleanup operations, nanomaterials collected through ventilation controls, and waste streams must be collected. To aid all this there should be process flow diagrams, facility schematics, and descriptions of the process with chemical reactions and standard operating procedures.

(ii) *Task Information on the Workforce:* The division of labor, the frequency of occurrence of tasks of workers, and use of personal protective equipment (PPE) needs to be included in this process.

(iii) *For Proper Control:* Sources of information should include plant rosters and organizational charts, job task descriptions, current job safety analysis, interviews with supervisors and workers, and detailed workplace observations.

Review of Safety Data Sheets (SDSs): Refers to the process of examining and analyzing previous sampling data on whether specific processes lead to direct airborne nanomaterial releases from, for example, vapor-phase synthesis reactors, heavy conveying or bagging operations, or shaping and grinding steps, which are those operations that take place in closed systems. Exposure potential can be higher when products are being transported or dried, during reactor maintenance and cleaning operations, material handling tasks like bagging or if the airborne nanoparticles are getting agglomerated in the coarse size range [11], since an aggregate may have a biologically relevant nanostructure. Peters *et al.* [12] have nicely assessed the concentration mapping that involves the measurement of particle concentrations by different metrics. hen developing methodologies to establish occupational exposure limits (OELs) for nanoparticles, a distinction needs to be made between "background particles" and "incidental nanoparticles." Different strategies are employed to develop methodologies to establish OELs, which include:

(i) *Hazard Assessment:* This is done by collecting available information based on the work done on the intrinsic properties of the substance, including physicochemical, toxicological, ecotoxicological, and fate properties. Criteria as to what aerosol fraction and what exposure metric is most relevant to health is established so as to derive the threshold levels. Assessing the hazards of NPs requires studying or knowing the physicochemical, toxicological, and ecotoxicological properties of the substance, in whatever size, shape or physical state and accordingly define their levels of exposure under reasonable conditions of use. It is also imperative to characterize the risks by comparing the levels of exposure and threshold levels below which risks to human health and the environment are considered to be controlled.

(ii) *Exposure Assessment:* This assesses short-term or long-term exposure to determine the exposure level to engineered nano material (ENM) by taking into account human target groups, routes of exposure, and environmental compartments. This will include production or R&D workers, consumers, and man *via* the environment scenarios such as pelagic sediments in both freshwater and marine environments, aquatic food chain, sewage treatment, indoor and outdoor air.

(iii) *Risk Characterization:* This can be achieved with the help of hazard assessment and exposure assessment. We have already discussed concerns about PM 2.5 in evaluating traditional methods of assessment of risks due to chemicals and dust. Health risks from nanoparticles share some features of the PM 2.5 risk calculations. However, since fundamental mechanisms for health risks from nanoparticles are not very well known and understood, not enough epidemiological data are available for engineered nanoparticle exposures and relevant health effect; it may be difficult to characterize those from PM 2.5. This is especially true because the technology is moving at a rapid pace and new nanoparticles are routinely manufactured. Thus, there is a need for quantitative frameworks that help in risk assessment and expert judgment. Hence, keeping track of the risk assessments presented in diverse scientific opinions about this emerging and metastable science of nanotechnology will be very useful. Moreover, the dose-dependent toxicity data and quantitative risk assessment contained in the exposure profile of each nanomaterial may differ. For risk assessment, if human dose-response data are not available then toxicology data from experimental studies in animals/rats on pulmonary inflammation and fibrosis are frequently used [13]. Moreover, granuloma, KRAS oncogene mutations, cardiovascular oxidative stress, and plaque formation, and cell proliferation in lung epithelial cells are also taken for base data.

3.5.1 Control Banding to Analyze and Prioritize Risks from Nanomaterials

Nowadays, control banding (CB) strategies for risk assessment of nanoparticles is being given preference [16, 17] because it offers simplified solutions for controlling worker exposure to constituents often found in the workplace. Although the original CB model was developed within the pharmaceutical industry, the simplicity afforded by CB can be particularly useful when dealing with nanomaterials; because (i) it minimizes or eliminates the need for complex quantitative assessments of worker exposure for comparison against occupational exposure standards, instead (ii) it provides a specific risk band based on a substance's hazard that is based on safety data sheets (SDSs), and (iii) it provides potential exposure, i.e., dispersibility characteristics.

The problem with nanoparticles is that they lack a clear toxicologic basis for setting OEL, due to their high degree of reactivity, ability to deposit in various regions of the respiratory tract, and their capability of crossing the blood-brain barrier (BBB) [14, 15] and even intact skin. The number of CB strategies for nanomaterials include: CB Nanotool, Stoffenmanager Nano 1.0, French Agency for Food, Environmental and Occupational Health & Safety (ANSES), etc., which varies depending on users and work area applications, such as research laboratories, medium-size companies, and small-size enterprises. All these need consistent inputs by experts to fill knowledge gaps.

A decade ago, in 2014 ISO issued a new technical specification standard on the use of CB for managing inhalation risk from engineered nanomaterials. The guidelines based on a CB approach are specifically designed to control the nanomaterials and their aggregates and agglomerates greater than 100 nm, that can be used as a first approach in the absence of relevant regulatory specifications.

3.5.2 How to Measure and Assess Exposure to Nanoparticles

Exposure measurement methods of nanoparticles incorporate detection of the presence of nanoparticles and their specific properties, whether they are present in air, water, soil, and consumer products, the media in which humans and ecosystems are exposed to nanoparticles. Though detecting nanoparticles is complex, both in gases and in liquids, recently some methods have become available for this purpose, so that even surface coatings may now be detected. To assess the exposure to nanoparticles there is a dependence on instruments. Hence, there is a need to develop advanced equipment to determine aerosolized and water exposures to engineered nanomaterials; along with developing techniques and approaches to assess the hazards of engineered nanomaterials and methods for estimating the environmental, health, and safety impacts (EHS) of nanomaterial exposures over a lifespan. This will facilitate the methodologies to strategically assess programs that could be utilized to implement relevant risk-focused research. Additional tools are required to detect nanoparticles in cells, fluids, or plant tissues.

3.5.2.1 Measurement of Airborne or Gaseous Nanoparticles

We are unknowingly exposed to the nanoparticles in ambient air, mainly from diesel fumes that release about 10-nm-diameter nanoparticles, which rapidly join together to form up to 100-nm-diameter nanoparticles that may remain in the air for days or weeks. Though exposure is low in terms of mass but significant in terms of the number of particles, needless to say, they reach lungs. Detecting and collecting nanoparticles samples in gas is rather complex and challenging, because huge volumes of air is to be collected for analysis and then separation of nanoparticles from larger particles can only be achieved under certain conditions. In addition, it is important to distinguish between existing ambient particles and engineered nanoparticles. Recently, an alternative procedure was developed involving the use of a mass spectrometer allowed for very precise measuring of the chemical compositions of nanoparticles of specific sizes inside a gas. There is a need to define which parameter—mass, number of particles, and/or surface area—should be measured as the most appropriate measure of assessing exposure. Most of the available instruments for nanoparticle exposure is inadequate, heavy, and not portable. There is a need for portable instrumentation and new sampling techniques and strategies for assessing exposure at the workplace and in the environment.

3.5.2.2 Assessment of Nanoparticles in Liquid

For measuring in liquids, there are commercially available instruments that can detect nanoparticles down to the 3 nm limit. However, for measuring nanoparticles in liquids a complicated sample preparation for analysis is required. At present, electron microscopy is used to study particle size, shape, and structure in liquids. When equipped with a specific spectrometer, the chemical composition can also be determined, at least for large nanoparticles. It is the number of particles, their size, and their surface characteristics that are determinants for interactions with living systems.

3.5.3 Methods Used to Assess the Risk Due to Nanoparticles

Hazard and risk assessment strategies for nanoparticle exposure has been envisaged, considered, and evaluated for more than a decade now. Since then, this applied science is maturing and commercially flourishing. Many newly developed products are in the process of being adequately evaluated in the European Union through European Commission FP7-sponsored programs on nanosafety issues; and by the United States Academy of

Science Committee on Research Progress of Environmental, Health and Safety Aspects of Engineered Nanomaterials. Some NGOs related to the Environmental Defense Fund and companies synthesizing and commercializing nanoscale products are also fostering the NanoRisk framework effort.

So far as the hazardous impact on human health is concerned, there is a realization that lung exposure to nanoparticles is more toxic than fine-sized particles of the same material. Two other major routes of exposure, i.e., ingestion (oral) and dermal, have also been raised. Hence, there is a need to review the conceptual planning in use, and consider the shortcomings and revise the implementation strategies and activities. An adequate investigation regarding health effects is suggested for (i) different nanomaterial variations within a specific nanoparticle, e.g., carbon displays many different nanoscale forms (fullerenes, single-walled carbon nanotubes, multi-walled carbon nanotubes, carbon nanofibers, carbon dots, carbon spheres, carbon black particles); (ii) various possible routes of exposure; (iii) nanomaterial characterization issues; (iv) limitations in research methodologies such as time-course and dose-response issues, and (v) guideline for toxicity testing of *in-vitro* methodologies for *in-vivo* standardization.

Considering these aspects, a dilemma arises regarding the testing of each individual nanoparticle, as this will require overwhelming expense, time required for short- as well as longer-term exposures to humans or in the environment. Although there have been considerations on the "conceptual" and organizational front, the challenges of assessing the health effects of specific nanoparticle types remain a daunting task for better estimation of potential adverse health effects following nanoparticle exposure. However, we have reached a consensus that the following processes should be considered critical for future research endeavors:

(i) Prior to toxicity studies, thorough characterization of the material should be done using standardized protocols
(ii) Using established processes for validation of test methods and toxicity results; round-robin interlaboratory studies should be carried out. A round-robin study is an interlaboratory experimental methodology that helps to determine reproducibility of a "process" where tests are performed independently multiple times and the results are analyzed statistically to assess their variability.
(iii) It is better to carry out protocols and standardizing methodologies in governmental organizations, funded by government; to test commercially relevant nanoparticle types (to which humans will be exposed).

(iv) It is critically important to gauge medium- and/or long-term effects of relevant nanoparticle exposures at relevant dose levels for both *in-vitro* and *in-vivo* studies. It is necessary to develop an effective relevant protocol to transition from *in-vivo* effects to *in-vitro* effects. For this, a selection of relevant cell types and cell models that reflect the routes of human exposure should be chosen and validated by different research groups. This will help in developing *in vitro* toxicity methodologies or cell culture techniques to better simulate longer-term effects, including more realistic dosimetry and better particle characterization techniques.

(v) There are regulatory standpoints for hazards of fine particles. However, the question of whether nanoparticles are more hazardous than fine particles of similar chemistry has not been adequately addressed. Moreover, from a regulatory standpoint, more useful definitions for defining nanomaterials are necessary.

All these considerations demand more emphasis on focusing on the research related to biokinetics/toxicokinetics of nanoparticles at human exposure levels. This is especially so when considering the pulmonary, oral, and dermal routes of different nanoparticle exposures.

3.6 Are Sufficient Methods Being Used to Assess the Risk from Nanoparticles?

Adequate methodologies for assessing the risk from nanoparticles have to have specified considerations of factors of exposure, how to address the risk and hazards related to nanoparticles, and what should be done to improve their risk assessment. Moreover, since chemicals in their nanoparticle form exhibit different properties from their larger physical forms, they may interact differently with the biological systems and the environment. Therefore, conventional toxicity tests may not be enough to detect all their possible harmful effects. There are a series of specific tests being proposed to assess the toxicity of nanoparticles used in drug delivery systems; for example, toxicity of nanoparticles is likely due to the induction of oxidative stress in cells and organs. Hence, testing for the interaction of nanoparticles with proteins and various cell types should be considered as part of the toxicological evaluation. So far as the status of nanomaterial safety efforts is concerned, it seems that many challenges yet remain in providing accurate health and environmental decisions on many of the nanomaterials. Significant improvements have to be made in the near future. To answer the question of whether the methods being used are sufficient to assess the risk, the following points should be looked into:

(i) The biological properties of only a few types of nanoparticles have been studied, and chances are that they may not be representative of nanoparticles in general. At the moment, there are no official guidelines on risk assessment methodology regarding exposure assessment, hazard identification, hazard characterization, and risk characterization. Because the commercial production of nanoparticles of different types is continually evolving and relatively new, there is very little data available on their effects on human populations or the environment.

The following nanoparticle safety requirements need to be addressed: (a) Safety of workers involved in production and transport during the manufacture of nanoparticles, and of consumers using products that contain nanoparticles; (b) safety of local human and animal populations due to the release of nanoparticles from manufacturing and/or processing facilities; (c) impact of nanoparticles on the environment and the potential for human re-exposure through the environment; and (d) environmental and human health risks involved in the disposal or recycling of nanomaterials.

Above all, more safety requirements should be developed for new or emerging nanotechnologies; than for tried and tested technologies. If these expectations are not met, chances are that the public will fear and even reject products based on nanotechnology.

3.6.1 Considerations to Assess Risks that Should be Taken into Account

The following considerations are being envisaged for nanoparticle risk assessment:

- Human exposure to nanoparticles can be through inhalation, ingestion, dermal contact and injection, and implantation.
- Nanoparticles may enter through cell membranes to reach the cells of several organs, and they can even cross the blood-brain barrier (BBB) to enter the brain. The ability to pass through cell membranes may either increase gradually as particle size decreases or increase suddenly below a certain particle size.

- The adverse effect of nanoparticle has not been observed with other chemicals in larger physical forms. Therefore, nanoparticles may have greater toxicity or, perhaps, different toxicity compared to larger particles of the same substance.
- Nanoparticles may have the potential for bioaccumulation in humans and possibly in other species and the environment. Therefore, the hazardous exposure dose should be measured in terms of mass total surface area, number of particles, and their combination.
- So far as the impact of nanoparticles in the ambient air is concerned, it will probably be widely dispersed and they may distribute and persist in the environment differently from larger particles. Hence, the potential problems associated with nanoparticles that persist in the environment may be considerable.
- People with severe chronic respiratory and heart diseases are liable to be more sensitive to the adverse effects of certain nanoparticles from ambient air pollution than the general public, and potentially to other types of airborne nanoparticles.

3.6.2 Exposure Factors Required to be Specified for Risk Assessment Methodologies

Deciding factors for exposure in a risk assessment will depend on either a chemical in nanoparticle form or to the product in which these nanoparticles are incorporated. For both of these the physical and chemical properties, solubility, bioaccumulation, biodegradability, the amount produced, their expected uses, and proposed routes for disposal or recycling, will be the major factors for consideration. Exposure to both humans and the environment, as well as the potential for accumulation in the body after repeated exposure, including the amount, distribution, and persistence of nanoparticles in the environment, as well as the potential for bioaccumulation in different species, should be given attention. Nanoparticles can cause hazards mainly due to the following factors:

(i) The toxicological properties of the chemical composing the nanoparticle are expected to have properties similar to the chemicals in larger, conventional forms, unless the distribution of the nanoparticles in the body or the environment is very different.

(ii) Reactivity caused by the nanoparticles due to their large surface area to volume ratio will have toxic effects different from those of the same chemical in larger physical forms.
(iii) The potential of nanoparticles to adsorb other chemicals of concern onto their surfaces; as well as the potential for the release of these adsorbed chemicals of concern will also need to be addressed.
(iv) A combination of all the above factors.

Finally, it is to be stressed that each product and process involving nanoparticles has to be considered separately when assessing the safety of workers involved in its manufacture, of consumers and the public living near manufacturing or processing facilities, the environmental impact, and the human health and environmental risks involved in the disposal or recycling of the nanoparticle products.

3.6.3 How to Improve the Risk Assessment of Nanoparticles?

To improve the risk assessment of nanoparticles, critical gaps in knowledge need to be addressed. This can be done through the new techniques for the measurement of exposure levels of nanoparticles in humans and the environment, new protocols to assess the release of nanoparticles from various products and production processes, and by extrapolating from the toxicology of non-nano materials to nanoparticles of the same substance, and between nanoparticles of different size ranges, etc. Moreover, by compiling information on the health of workers involved in the manufacture and processing of nanoparticles, and information on the effect of nanoparticles in the environment on various species as well as current and past exposure of humans and the environment to nanoparticles data collection.

3.7 Conclusion

Prior to making a concluding remark, it is worth mentioning the efforts of REACHnano consortium, who in 2015 proposed "Guidance on available methods for risk assessment of nanomaterials." This was an outcome of the joint effort of the European Union's LIFE fund in the framework of the LIFE programme environmental policy and governance entitled "LIFE REACHnano" aimed at developing a web-based REACH toolkit to support chemical safety assessment of nanomaterials (LIFE11 NV/ES/000549).

However, with the development of new nanomaterials of different configurations, the safe and responsible use of the technology remains a high priority to be further evolved with greater sophistication. Because existing methodologies to assess the potential human health and environmental risks associated with products of nanotechnology are not appropriate, it must be accepted that even if the existing methods are appropriate to assess many of the hazards associated with the products and processes involving nanoparticles, they may not be sufficient to address all the hazards. Especially since the dose of exposure in terms of mass alone is not sufficient; it also needs to be expressed in terms of total surface area, several particles, or a combination of the two. The mode of delivery of the nanoparticle to the test system needs additional exposure assessment for environmental hazards. For this to happen, existing methodologies need to be modified as well as new ones developed to determine the physical and chemical properties of nanoparticles routinely.

The uncertainties introduced by nanotechnology has given rise to some questions, including:

(i) What is the amount of time that manufactured NPs can persist in the atmosphere?
(ii) What is the stability of NPs over time given specific conditions?
(iii) What is the effect of shape and size on their fate and transport?
(iv) What are the likely routes of exposure (e.g., inhalation, dermal, ingestion, or ocular)?
(v) What are the metrics to measure the exposure of NPs (e.g., particle mass or number or surface area, concentration)?
(vi) What are the key mechanisms of translocation to different parts of the body?
(vii) What are the possible mechanisms of toxicity and oxidative stress due to surface reactivity, presence of transition metals leading to intracellular calcium and gene activation, and intracellular transport of NPs to the mitochondria [19]?

Simultaneously, record the mechanisms, actual levels of and rate at which nanoparticles are released from products and processes affecting the exposure and related health effects of workers involved in the production and processing of nanoparticles. Except for airborne particles delivered to the lung, information on the behavior of nanoparticles in the body, including distribution, accumulation, metabolism, and organ-specific toxicity is still minimal. Moreover, data are needed on the effects of nanoparticles on various environmental species, movement of nanoparticles inside the body and their accumulation in different species, including microorganisms. Also, the behavior of nanoparticles in the environment, and how they are distributed if they persist in it.

References

1. Mohammad, A. and Młynarz, P., The toxicity of nanoparticles and their interaction with cells: An *in vitro* metabolomic perspective. *Nanoscale Adv.*, 5, 2674–2723, 2023, DOI: 10.1039/D2NA00534D (Review Article).
2. Mekonnen, G., Review on Application of Nanotechnology in Animal Health and Production. *J. Nanomed. Nanotechnol.*, 12, 3, 2021, DOI: 10.35248/2157-7439.21.12.559.
3. Shende, S.S., *et al.*, Interaction of Nanoparticles with Microbes, Plant-Microbes-Engineered Nano- particles (PM-ENPs) Nexus in Agro-Ecosystems, in: *Advances in Science, Technology & Innovation*, P. Singh, R. Singh, P. Verma, R. Bhadouria, A. Kumar, M. Kaushik (Eds.), Springer, Cham, 2021, https://doi.org/10.1007/978-3-030-66956-0_12.
4. Sonwani, S., Madaan, S., Arora, J., Suryanarayan, S., Rangra, D., Mongia, N., Vats, T., Saxena, P., Inhalation Exposure to Atmospheric Nanoparticles and Its Associated Impacts on Human Health: A Review. *Front. Sustain. Cities*, 3, 2021, https://www.frontiersin.org/articles/10.3389/frsc.2021.690444 ISSN= 2624-9634 DOI=10.3389/frsc.2021.690444.
5. Jani, P.U., McCarthy, D.E., Florence, A.T., Titanium dioxide (rutile) particle uptake from the rat GI tract and translocation to systemic organs after oral administration. *Int. J. Pharm.*, 105, 2, 157–168, 1994.
6. Gautam, A., Singh, D., Vijayaraghavan, R., Dermal exposure of nanoparticles: An understanding. *J. Cell Tissue Res.*, 11, 1, 2703–2708, 2011, ISSN: 0974-0910.
7. Karin, A., Micheletti, C., Sokull-Klüttgen, B., Christensen, F.M., Analysis of currently available data for characterising the risk of engineered nanomaterials to the environment and human health — Lessons learned from four case studies. *Environ. Int.*, 37, 6, 1143–1156, 2011, ISSN 0160-4120, https://doi.org/10.1016/j.envint.2011.02.005.
8. ECHA, https://echa.europa.eu › Doc.: MB/17/2009 final. General Report of the. European Chemicals Agency. 2008. 24 April 2009. (Art. 78(a), 83(3) of Regulation (EC) No. 1907/2006).
9. Giannadaki, D., Lelieveld, J., Pozzer, A., Implementing the US air quality standard for PM 2.5 worldwide can prevent millions of premature deaths per year. *Environ. Health*, 15, 88, 2016. https://doi.org/10.1186/s12940-016-0170-8.
10. Kandlikar, M.,.G., Ramachandran,.A., Murdock, M.B., Health Risk assessment for nanoparticles: A case using expert judgement. *J. Nanopart. Res.*, 9, 137–156, 2007, DOI 10.1007/s11o51-006-9154-x.

11. Peters, T.M., Elzey, S., Johnson, R., Park, H., Grassian, V.H., Maher, T., *et al.*, Airborne monitoring to distinguish engineered nanomaterials from incidental particles for environmental health and safety. *J. Occup. Environ. Hyg.*, 6, 2, 73–81, 2009.
12. Peters, T.M., Ott, D., O'Shaughnessy, P.T., Comparison of the Grimm 1.108 and 1.109 Portable Aerosol Spectrometer to the TSI 3321 Aerodynamic Particle Sizer for Dry Particles. *Ann. Occup. Hyg.*, 50, 843–850, 2006.
13. Nakanishi, J. (Ed.), Risk assessment of manufactured nanomaterials: carbon Nanotubes (CNT). Final report issued on August 12, 2011. NEDO project (P06041). New Energy and Industrial Technology Development Organization, Kawasaki, Japan, http://en.aist-riss.jp/ 2011.
14. Sharon, M., Should Nano-particles Be Used to Cross Blood-Brain Barrier to Deliver Drugs to Neurodegenerative Disease? A Mini-Review. *Nov. Appro. Drug Des. Dev.*, 1, 1, 555554, 2017.
15. Chinmay, P., Dharmatti, R., Sharon, C., Mewada, A., Mugdha Bedekar, Sharon, M., Azadirachta indica (Neem) Gum Coated Gold Nanoparticles as Nano-go-karts to Dispatch Haloperidol across Blood-Brain-Barrier. *Int. J. Pharm. Sci. Rev. Res.*, 38, 2, 167–172, 2016.
16. Brouwer, D.H., Control banding approaches for nanomaterials. *Ann. Occup. Hyg.*, 56, 5, 506–514, 2012, 2012.
17. Kuempel, E.D., Castranova, V., Geraci, C.L., Schulte, P.A., Development of risk-based nanomaterial groups for occupational exposure control. *J. Nanopart. Res.*, 14, 9, 1–15, 2012.
18. Ruoyu, Z., Lu, X., Wang, L., Chen, J., Chen, D.-R., Wang, X., Extremely Inexpensive and Simple Method to Remove Indoor Respiratory Aerosols. *Environ. Sci. Technol. Lett.*, 2023. https://doi.org/10.1021/acs.estlett.3c00541.
19. Kandlikar, M., Ramachandran, G., Maynard, A., Murdock, B., Toscano, W.A., Health risk assessment for nanoparticles: a case for using expert judgment. *J. Nanopart. Res.*, 9, 1, 137–156, 2007.

4

Human Health Risks Due to Inhaled Nanoparticles

If you believe in science, as I do, you believe that certain laws are always obeyed.

Stephen Hawking

Abstract
Inhalation is a major route of nanoparticle exposure due to their small size. This chapter explores both natural and man-made sources of airborne nanoparticles and their journey through the respiratory system into other organs. Once in the body, nanoparticles can trigger oxidative stress, inflammation, genotoxicity, and tumor formation. The chapter discusses health risks to the lungs, cardiovascular system, digestive tract, kidneys, brain, reproductive organs, and endocrine system. It also considers the potential of nanoparticles in cancer therapy, presents bioassay methods for toxicity assessment, and evaluates current risk analysis approaches.

***Keywords*:** Atmospheric-nanoparticle, genotoxic, tumorigenic, ultrafine, respirable, inhalable, natural-source, anthropogenic-source, inhalation-route

4.1 Introduction

This chapter focuses on the atmospheric nanoparticles (NPs); especially their source, hazardous impact on different human body systems leading to diseases when inhaled through the respiratory system. And the toxicological, genotoxic, and tumorigenic effects in humans caused by exposure to NPs, the origin of atmospheric NPs, and their transportation, deposition, and toxicity.

Due to a lack of data, there is deep scientific uncertainty regarding what happens once the NP is inhaled; and its further mechanisms of translocation to different parts of the body, routes to different organs in the body, and the mechanisms of toxicity and disease caused. The metrics by which

Madhuri Sharon. Risk, Safety, and Ethical Concerns in Nanotechnology, (63–88) © 2025 Scrivener Publishing LLC

exposure ought to be measured is another concern to be standardized. To understand the risk caused by inhaled particle, PM 2.5 has emerged as a priority aspect. The particle sizes being expressed in common language are:

- Ultrafine are NP of <100 nm
- Fine >0.1 μm–3 μm
- Respirable (by rat) <3 μm
- Respirable (by human) <5 μm
- Inhalable (by human) ~10–20 μm.

The fate and transport of larger particles is well understood and inhalation is the exposure route. Research efforts have shown that there is substantial new information about the early/short-term biological impact post exposure; however, long-term health effects and the mechanisms of these effects are poorly understood.

4.1.1 Sources of Nanoparticles

Nanoparticles (NPs) production is increasing day by day for their use, due to their unique properties. Moreover, there are also anthropogenic and natural sources of NPs. They cause air pollution due to their ultrafine size, and can remain suspended in the atmosphere for a long time, can travel larger distances. A variety of NPs are found in indoor and outdoor settings and cause several health issues after exposure. All of them, when inhaled or through active and passive smoking, cause respiratory malfunctioning, further leading to various diseases related to different organs. The NPs translocate to the central nervous system either through neural pathways in the olfactory system or through the blood and lymphatic circulatory systems (blood) after being absorbed in the lungs, and eventually may cross the blood-brain barrier and enter the brain. Whereas, when larger particles in the PM range are inhaled, they are deposited in lungs and cause localized damage to the pulmonary system, as they are not accessible to or cannot be translocated to parts of the body. There are chemically different types of NPs, depending on the source of their origin.

4.1.1.1 Natural Sources of Nanoparticles that we Cannot Avoid

Nature has abundant NPs. Source of these NPs are biogenic emission, sea spray, forest fire, volcanic eruption, landslide, and dust storm. Moreover, naturally generated aerosols are also present in the atmosphere, as well as generated by human activity. Similarly, organic compounds are also in the atmosphere in large amounts that can react with clouds and air particles [64]. Emission sources of natural nanoparticles are from (i) biochemical recycling of bacteria, fungi, and plants; (ii) abiotic chemical precipitation such as by sea spray; (iii) incomplete combustion of forest fires; (iv) geologic sources like volcanic eruption, earth quake, dust storm, and glaciations. Ashes of volcanic eruptions comprise both solid and liquid PM, having high concentrations of Ni, Zn, Cd, Ag, Sn, Se, Te, Hg, Tl, Pb, and Bi NPs [14]. Moreover, NPs composed of OC are one of the major air pollutants. There are some analyses of NPs at different locations caused by various reasons [32, 53, 70].

4.1.1.2 Anthropogenic Sources of NPs that Need Careful Management

Emission sources of anthropogenic nanoparticles could be intentional or unintentional; for example:

(i) *Unintentional Sources:* These are mostly from automobile and industries that release incomplete burning output; also, biomass, and non-biodegradable garbage burning. Exhaust from vehicles, ships, submarines, airplanes, rocket launches, etc., are some dangerous mobile sources of NPs [77, 78]. The NPs of exhaust include carbon monoxide (CO), hydrocarbons (HCs), nitrogen oxides (NOx), sulfur oxides (SOx), volatile organic compounds (VOCs), and particulate matter (PM) [3, 84]. Analysis of vehicle exhaust by Buzea *et al.* [7] has shown that they are in the size range of 20–60 nm for petrol engines and 20–130 nm for diesel engines.

(ii) *Intentional Sources:* These are increasing day by day with the use of different types of mobile vehicles, pesticides, herbicides, fertilizers, chemical industries, metallurgical industries, thermal power plants, energy production, ore extraction, mining, excavation, and construction; as well as dust cloud caused by demolition that sometimes rises very high into the atmosphere, affecting a larger area; and from water and solid remediation systems. Domestic fuel combustion smoke from inefficiently burning stoves and use of biomass fuel also adds up to indoor generation of NPs. Cigarette smoke that emits approximately 8.8×10^9 NPs per cigarette, is also a common source of NPs; whether it is active or passive smoking. There should be strict regulations to prevent or control the intentional sources of NP release into the environment.

4.2 Inhalation Routes of NPs from Nose and Lungs to Other Organs of Human Body

Nanoparticles (NPs) are inhaled through both nasal and oral routes (Figure 4.1). After passing through the larynx, pharynx, and trachea, bronchial bifurcations reach the alveoli, which are the tiny sacs in the lungs. Lungs, as we know, are the part of the respiratory system made up of a group of tissues. The lung is an organ for purification of metabolically used blood, that removes carbon dioxide from blood and supplies oxygen taken from inhaled air; and then transports the purified blood to all the organs of the body. There are about 3,00,106 alveoli [7] in lungs. It is the alveoli which is the workhorse for purification of blood. Once NPs enter the alveoli, being very tiny sacs, they can retain as well as easily transport the NPs through the alveolar epithelium and capillary endothelial cells to enter the cardiovascular system and other internal organs [61]. The NPs can penetrate outer as well as inner cellular compartments, into the cytoplasm and karyoplasm

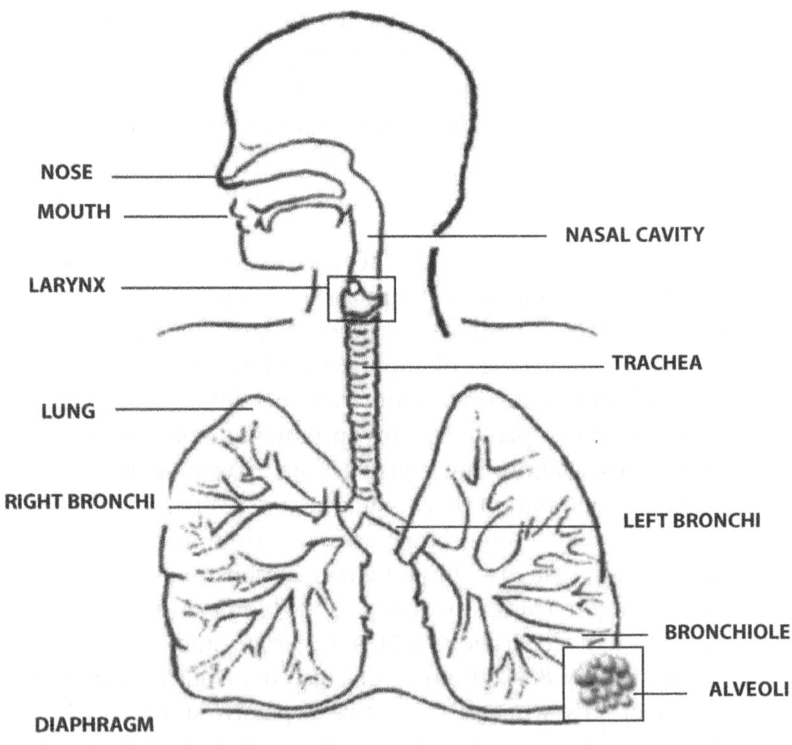

Figure 4.1 The human respiratory system.

of pulmonary epithelial and mesothelial cells of the lungs [2]. Metal fumes containing Cu and Zn, and metal dust of Ni, Cr, and Co are known to cause pulmonary fibrosis and in severe cases, cancer [7]. The deposition of NP in respiratory tract takes place by Brownian movements. Whereas according to Ferreira *et al.* [16], the site of deposition is governed by the aerodynamic diameter, and NPs (≤20 nm diameter) have a high probability of reaching the alveolar region. Geiser *et al.* [18] found that within 24 h of inhalation, 20 nm aerosol NPs pass through the respiratory tract and get deposited in the alveolar region of the lungs.

4.2.1 Translocation of NPs to Different Sites of the Body

Unlike inhaled larger particles, the NPs easily get translocated to different parts of the body and due to their small size are taken up into the cells. The extent of translocation of NPs may on their chemical characteristics. Inhaled NPs have been detected in the interstitium of different organs (interstitium is a contiguous fluid-filled space existing between a structural barrier, such as a cell membrane or the skin, and internal structures such as kidneys, muscles and the circulatory system) [48], in the lymph nodes [49], and circulatory system [44, 31]; and are transported through neuronal pathways to the central nervous system and the brain [50]. The NPs circulating in the blood and lymphatic system may be translocated to bone marrow, spleen, and heart. The insoluble iridium NPs deposited in lung epithelium have been shown to translocate to extrapulmonary organs; however, the number of NPs is very low and size-dependent, and within 7 days after inhalation they were cleared *via* airways into the gastrointestinal tract and feces. This may induce adverse reactions in the cardiovascular system and liver [31].

4.3 Cytotoxicological Considerations

4.3.1 Cellular Response

There are four main types of cytotoxicological effects of inhaled NPs at the cellular level, which are discussed below.

4.3.1.1 Oxidative Stress

The first response to NPs is oxidative stress leading to nanotoxicity. It is caused by the imbalance that is created between free radicals and antioxidants in the body. The imbalance in the respiratory system occurs due to the extra free radicals with an uneven number of electrons which react inadvertently with other molecules [75].

4.3.1.2 Inflammation

Our body defends itself against any foreign particle (called an antigen) entering it. Even if particles as small as NPs enter the body through an innate immune system, an antibody will deactivate it. However, if the antibody fails to deactivate the foreign particle, a complex adaptive immune system is induced. There are dendritic cells in the body, which are known to carry out this activation. Since they are foreign particles, NPs activate dendritic cells, which in turn generate ROS, cytokines, and chemokines and activate T-cells and various inflammasomes. The NPs elicit an inflammatory response in the respiratory and other systems of the body [9, 41, 52].

4.3.1.3 Genotoxicity

Nanoparticles are genotoxic substance that can induce genotoxicity either by their direct interation with DNA (This happens when NPs enter the nucleus and cause modification of DNA); or Indirectly because of oxidative stress or inflammation induced by NPs [93].

4.3.1.4 Tumorigenicity

Tumorigenicity is the tendency or the ability to produce or develop carcinogenic or benign tumors in body. Apart from their chemical composition, shape and particle aging, the quantum properties of nanoparticles, such as their minute size, are key factors in inducing tumorigenicity. Reactive oxygen species (ROS), and sometimes direct NPs, interact with the receptor on the cell surface, called the epidermal growth factor receptor (EGFR), which results in the formation of abnormal proteins [76].

4.4 Known Hazardous Impact of Inhaled Nanoparticles on Human Health

The initial passage of inhaled NPs into the alveoli of lungs is one of the major pathways of NPs in the human body. The NPs deposited in alveolus (air sacs), largely escape the alveolar macrophage surveillance and gain access to the pulmonary interstitium. The lung interstitium is the space between the air sacs and the capillary blood vessels that surround them. Through endocytosis, NPs in the alveolar epithelial cells are transported *via* blood to the other organs of the body. The other pathway involves the diffusion of NPs into the olfactory bulbs of neurons *via* the nasal epithelial barrier to reach the central nervous system, which is spread throughout the body, and therefore the NPs get translocated to the different organs of the body [13, 26].

One of the prominent effects of inhaled NPs is that it elicits an Inflammatory response that can affect many organs. It is detected by the increase in the number of neutrophils and in the level of cytokine and chemokine, which is due to the oxidative stress pathways of the cells [20, 67, 68]. There has been a lot of research work done on the inflammatory response to the PM, based on which it is envisaged that there may be systemic inflammatory responses characterized by circulating neutrophils, fibrinogen, plasma viscosity, platelet numbers, and C-reactive protein [25, 54, 56, 67, 72, 73, 83]. Almost all the organs of the human body are impacted by NPs.

4.4.1 The Respiratory Tract: Portal of Entry for NPs into the Human Body

The portal of entry is the means by which an infection is able to enter into the human body. Inhalation (*via* the respiratory tract) is the main gateway to body. However, NPs can also enter the human body by absorption *via* mucous membranes such as in the eyes; and ingestion *via* the gastrointestinal tract. The airway nerve or vagus nerve innervates the respiratory tract and regulates the airway functions such as breathing pattern and the Hering-Breuer inflation reflex, airway smooth muscle tone, mucus secretion, and cough. In response to NPs, it is envisaged that it may cause neurogenic inflammation where airway nerve cells synthesize neurotransmitters; affecting white blood cells and epithelial tissue [5]; this inflammation may cause reduced respiratory function or increase in susceptibility to infections. Whether it is active or passive smoking, it is reported that it causes thick septum, shortness of breath, and a high level of interleukin protein and tumor necrosis factor (TNF-a) that cause tumor generation in the exposed population. Bonner [27] had reported more than a decade ago that CNTs are a potential cause of pleural and interstitial lung disease and allergic asthma. The effect of NPs on the lung and respiratory system has been noted to be through Inflammation [9]; oxidative stress [75], genotoxicity [93], and tumorigenicity [79]. Entry of NPs into alveoli causes activation of macrophages, causing pro-inflammatory cytokine release [45].

Renwick *et al.* [63] have shown that due to the inflammatory response the epithelial cell layer and macrophages get damaged, causing decreased phagocytosis; hence, a decrease in clearing the NPs from the lungs occurs. Oberdorster *et al.* [49] studied the correlation between particle size and its persistence *in vivo* causing lung injury. They concluded that adverse pulmonary effects of NPs were due to their larger specific surface area, the greater interstitial access, and their altered biopersistence, hence increased retention; and suggested that pulmonary compartmental doses should be considered when evaluating effects (alveolar versus interstitial versus total dose) and the intrapulmonary kinetics and translocation of NPs.

Several studies have shown that while being retained in the lungs, the NPs have caused adverse effects, resulting in different diseases. Inhaled NPs or even NPs used for drug delivery are bound to get deposited in lungs. There they (i) affect the immune system, causing allergies and asthma; (ii) cause injury to lung tissue, ending in fibrosis, pleural fibrosis or mesothelioma; (iii) damage DNA, becoming a cause of lung cancer; and (iv) if (prior to clearance or even after clearance the NPs left behind) translocated to other organs many possible potential adverse effects are envisaged [27].

Two major health risks to lungs are:

(i) *Lung Tumorigenesis:* As we have seen above, the exposure to NPs results in generation of ROS, causing cytotoxicity, genotoxicity and eventually leading to tumorigenesis. Exposure to MWCNTs is an established lung tumor promoter. The tumorigenic potential of nano-quartz and TiO_2 NPs [86] and SWCNTs [87] in rat model; and CeO_2 and Fe_2O_3 was studied using pSAEC (primary small airway epithelial cell) [79] *in vitro*; both demonstrate that NPs have potent adverse effects in the form of inflammation and subsequent tumors in lungs.

(ii) *Pulmonary Fibrosis:* Inhaled metal fumes of copper and zinc NPs and dust of nickel, chromium, and cobalt cause pulmonary fibrosis and, in severe cases, even cancer [7]. It has also been reported that CNT/CNF nanorods, nanotubes and nanofibers are the cause of fibrosis.

It is interesting to report here the findings of Jaques and Kim [28] that total lung deposition fraction of ultrafine aerosols, regardless of particle size and breathing pattern, was greater in women than men.

4.4.2 Blood Circulation System: A Trojan Horse for Inhaled Nanoparticles

When inhaled or taken orally, nanoparticles (NPs) can directly enter the blood circulatory system from the lungs; or they can diffuse through the gastrointestinal tract and subsequently reach the circulatory system. Studies have been performed *in vivo* in which fluorescent NPs injected into rat lungs could be traced (near-IR imaging) to their entry into the bloodstream. The NPs smaller than 5 nm diameter and dipolar (both positively and negatively charged) traveled from the lungs to the lymph nodes and into the bloodstream within a few minutes, and shortly afterwards to the kidneys, suggesting that they may be on the way to getting expelled from the system or some other NPs may get deposited on their way to the kidneys or other organs.

The mechanisms responsible for the effects of NPs are not well elucidated, such as to what extent and how rapidly inhaled pollutant particles pass into the systemic circulation. Passage of inhaled NPs into the blood circulation system in human volunteers was assessed by Nemmar [44] using ^{99m}Technetium-labeled carbon particles (<100 nm). They found that radioactivity was already detected in blood at 1 minute, and by 20 minutes maximum radioactivity was recorded; it remained at this level up to 60 minutes and substantial radioactivity was noted over the liver and other areas of the body. The results showed that ^{99m}Tc-carbon NPs diffuse rapidly into the systemic circulation, suggesting that this should be considered relevant for the cardiovascular morbidity and mortality related to ambient particle pollution. In the blood circulatory system, NPs have been reported to increase the blood pressure, decrease heart rate, and alter vascular tone, leading to dysfunction of the cardiovascular system [90].

The blood circulatory system may not only act as cargo vehicle, but is bound to interact with the blood's components, i.e., erythrocytes, platelets, white cells, plasma and complement proteins; and can produce potential harmful effects [10] (Figure 4.2). So far as erythrocytes or red blood cells are concerned, their interaction with NPs varies with the intrinsic character of the NPs. For example, chitosan NPs are hematocompatible [82], whereas silica NPs caused hemolysis, deformation and aggregation of erythrocytes. Platelets, or thrombocytes, are small, colorless cell fragments in blood that form clots and prevent bleeding. Binding of silver NPs to platelet has caused the appearance of thrombin that can cause stroke or cerebrovascular accidents [33]. Lymphocytes or monocytes are peripheral mononuclear cells in the blood stream, which represent the host defense system. They were found to be adversely affected by the metal oxide NPs [38]. Nanoparticles have been found to cause the activation of the complement protein system, which is due to the surface charge.

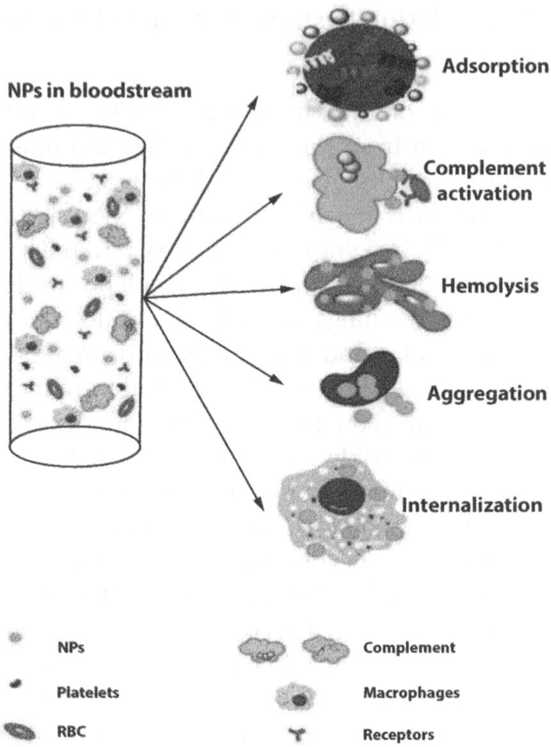

Figure 4.2 Interaction of nanoparticles with bloodstream components and their physiological effects. (Source: InTech. doi:10.5772/intechopen.69386).

4.4.3 Cardiovascular System Leading to Cardiopulmonary Disease

In nanomedicine, a variety of nanoparticles, such as lipids, polymers, biomimetic and silicon-based formulations, are extensively being used nowadays to deliver protein-, RNA- and small-molecule-based (including statins) therapies for the cardiac system to reduce oxidative stress, improve myriad heart functions, such as fractional shortening, increase wall thickness, etc., However, the adverse impact of NP on the heart as well as bloodstream has also been noted. The translocation of NPs takes place through the bloodstream; hence, the envisaged effect is expected to affect the ability of blood to coagulate [44]. Higher levels of fibrinogen could aid in increasing plasma viscosity and blood coagulation, leading to an increased formation of blood clots. Combustion-derived nanoparticulate induces the adverse vascular effects of diesel exhaust inhalation, causing endothelial lining inhibition of NO pathways and vasoconstriction [42]. Angiogenic/antiangiogenic, vasodilation/vasoconstriction, prooxidant/antioxidant, cytotoxic, apoptotic, and phagocytic effects of silver NPs have been found to depend on the physical properties, concentration, and retention time of NPs [21]. They have suggested that "this is because Silver NPs used in pacemakers, drugs, and coupled antibodies have a dual and opposite effect on blood composition, angiogenesis (vessel formation), and the permeability of membranes." Studies have indicated that systemic inflammatory response syndrome to SiO_2 NPs in old rats may affect atherosclerosis, arrhythmias, myocardial infarctions, increased blood pressure and decreased heart rate. Moreover, increased fibrinogen (FBG) concentrations and blood viscosity, along with myocardial ischemic damage and atrioventricular blockage is also observed [90].

4.4.4 Digestive System: Can it Digest Nanoparticles?

The digestive system consists of the gut, stomach, esophagus, small and large intestine. Guts mostly get exposed to NPs through water intake. However, NPs that get into the blood circulatory system *via* lungs are also prone to enter into the gut. Depending on the size, dose, dose duration and functional groups, NPs have been shown to adversely affect gut microbiota; however, there is a need for more studies and standardized protocols for characterization of NPs, dosing, and test systems [80]. Both the gut and stomach have been found to be affected by NPs, which causes inflammation of the stomach as well as high mucous production [19].

The NPs affect the gastrointestinal tract by modulating its overall homeostasis by altering the gut microbiota and gut-associated immune response. Different NPs behave differently in the gastrointestinal tract. Therefore, serious consideration is needed before using NPs, especially in the food industry to improve the quality, shelf life, safety, and nutritional profile of foods, as well as to enhance the functional performance of packaging materials, because accumulation of ingested NPs in the lamina propria of the intestines where they disrupt the function of goblet cells. Increased toxicity levels of NPs are expected to lead to colon cancer and other carcinomas. In a recent article by Zhou *et al.* [91] gastrointestinal fate of organic and inorganic NPs in Foods is discussed.

4.4.5 Kidneys: Can They Flux Out All NPs

Kidneys are an organ for blood filtration for the elimination of waste from the body. They are envisaged as an organ to transport and clear NPs from the body by glomerular filtration. Some nanoparticles that are being researched for use in nanomedicine have not been found to be renal clearable, such as liposome, though it has been cleared by the FDA to be used in Doxil® [4], polymeric micelles as a drug delivery vehicle [43], dendrimers [46], and magnetic iron NPs [62]. Whereas there are some renal clearable NPs such as gold NP [92], silica NP (C-dots) [57], and quantum dots [8].

Studies have suggested that nanoparticles exposed to the kidney could provoke glomerular swelling, basilar membrane thickening, degeneration, and necrosis of renal tubular cells.

In-vitro culture studies using HEK293 cells (human embryonic kidney cells) exposed to SiO_2 NPs, at dosage levels of 20–100 μg/ml, showed signs of apoptosis (shrinkage of cells and nuclear condensation). Moreover, ROS production in HEK293 cells and oxidative stress induction was also found, suggesting the nephrotoxic potential of SiO_2, which is a serious issue as it may lead to dysfunction in the elasticity of nephrons and cause cerebral epilepsy. A study on the cytotoxicity and oxidative stress induced by ZnO and CdS on glomerular and tubular human renal cells HK-2 cell lines of the lining indicated their nephrotoxic potential and epithelial lining cytotoxicity [60].

4.4.6 Central Nervous System and Brain without an NP Barrier

The central nervous system (CNS) encompasses the brain and spinal cord, which are made up of two basic types of cells: Neurons, the nerve cells that send and receive signals; and Glia, cells that provide structure in the brain. Capillaries that carry blood to the spinal cord and brain have a filtration system that does not allow unwanted material to enter the CNS such as pathogens, diffusion of solutes in the blood, and large or hydrophilic molecules into the cerebrospinal fluid; and is known as the blood-brain barrier (BBB), which allows the diffusion of hydrophobic molecules (O_2, CO_2, hormones) and small nonpolar molecules. The interaction of NPs with neurons has provided evidence of both negative and positive effects. Gold NP of 2.5 nm can pass the BBB [74]. Whereas, Huang *et al.* [23] showed that silver NPs can cross the cell membrane of mouse neuron cells. It increased IL-1β (a potent pro-inflammatory cytokine produced by cells of the innate immune system); CXCL13 (a major regulator of immune responses); MARCO (a class A scavenger receptor that is found on particular subsets of macrophages); and GSS (Gerstmann-Sträussler-Scheinker disease, which is a rare, neurodegenerative brain disorder caused by prion (proteins) that misfold in the brain, primarily in the cerebellum) for inflammation and oxidative stress. Silver NPs caused Aβ plaques deposition in neuron cells by inducing APP, and reducing NEP and LDLR for Aβ plaque deposition. Accumulation of NP has been found to affect gene expressions [89], cause apoptosis, cell cycle alterations and oxidative DNA damage [81], and also inflammation and hormone imbalance [15, 22]. A recent article by Liu [36] reported that silver NP (in the range of 18.3–120.4 nm) used in nasal spray in rat was found to accumulate in different regions of the brain; indicating a realistic risk to brain health resulting from the application of Ag-containing commercial products. There are also reports of other NPs that could pass the BBB such as Si, Cu, Al, C and metal oxides. The route of airborne 36 nm ^{13}C synthesized from graphitic rod by electric spark discharge in an argon atmosphere at a concentration of 160 $\mu g/m^3$, was shown in rats to be *via* olfactory mucosa of the nasopharyngeal region of the respiratory tract and subsequent translocation *via* axonal olfactory nerve to CNS [47].

4.4.7 Nanotoxicity against Reproductive System Organs

Nanoparticles can force their way through the guard tissues (epithelial cells, placenta, placental barrier, germ cells, Leydig cells, Sertoli cells, and blood tests) of the reproductive system [85]. Hence, they affect both ovary and testes as well as gametes through both oxidative stress and inflammation to which the male reproductive system are more sensitive. Increased oxidative stress causes cell death and poor spermatogenesis. Oxidative stress is reported to be the primary cause of 30–80% of infertility problems in males [6].

The *in-vitro* genotoxic effects of titanium dioxide nanoparticles (n-TiO_2) in human sperm cell were studied *in vitro*, and were reported to be due to inflammation and cytotoxicity in testis [69]. The TiO_2-NPs could cross the blood–testis barrier and induced inflammation, cytotoxicity, and gene expression changes that led to impairment of the male reproductive system. Data suggested that there was DNA damage in human sperm exposed *in vitro* to n-TiO_2 concentrations, and the putative role of reactive oxygen species (ROS) as mediators of n-TiO_2 genotoxicity. The NPs acted as endocrine disruptors, creating problems in the process of oogenesis, leading to hormone disbalance [24].

A study of the effect of molybdenum trioxide on the function of ovaries in rats has shown significant attenuation in the serum level of estrogen in the MoO_3 NPs; the luteinizing hormones and follicle-stimulating hormone were elevated while progesterone level change was insignificant. The weights of the right ovary and the uterine body decreased significantly in the exposed group. It was concluded that MoO_3 nanoparticle exposure could cause an imbalance of sex hormones and decrease in body and ovarian weights in the female rats [1].

Both male and female reproductive systems are negatively affected by a number of NPs. Moreover, since transmission of genetic and epigenetic information to subsequent generations is carried out by germ cells, which serve as a link between generations; the hazards in relation to reproductive system, need careful consideration of the long-term effects and toxicity of NPs.

4.4.8 Hormone or Endocrine System Nanotoxicity

The endocrine system is composed of various glands that release hormones into the bloodstream. The thyroid is one such butterfly-shaped glands, situated at the base of the neck, which produces two hormones: thyroxine (T4) and triiodothyronine (T3), which regulate the rate at which the body uses fats and carbohydrates. These hormones affect every cell in the body. It was reported by Jiang *et al.* [29] that magnetic NPs causes overproduction of T3 thyroid hormone and decreases thyroid-stimulating hormone. The effect of palladium NP on the endocrine reproductive system (hormone receptors) of female rats was noted, where it caused blockage of signal cascades due to overstimulation of hormones [34]. Nanoparticles have been found to affect endocrine-disrupting chemicals, resulting in cytotoxicity and an imbalance in production [24, 39]. Inhaled nanoparticle-rich diesel exhaust regulated testicular function in adult male rats by degrading effects on testes and kidney hormones [35]. Moreover, serum levels of alanine aminotransferase, alkaline phosphatase, lipid peroxidation and pathological lesion were elevated after sub-acute oral exposure to zinc oxide NP [75]. Many such studies are being done to show the dark side of inhaled NPs, especially on human beings. Here, I have presented only a few examples related to various organs.

4.4.9 Cancer: Are Nanoparticles Friends or Foes?

There are many reports about the theragnostic use of NP for cancer such as imaging, drug delivery, photothermal therapy, etc. NP-based therapeutics can induce tumor cell death and in turn increase neo-antigen release from this tumor. As such, there is no confirmed evidence that NPs are carcinogenic. But there is a report that NPs may speed cancer progression by speeding up the movement of cancer cells from the original tumor site to new sites and help the cancer cells that are already in motion evade blood circulation [55].

However, prolonged exposure to inhaled overburdening of NPs in lungs due to their intrinsic reactivity and processes such as inflammation may cause carcinogenic changes. When NPs are used in cancer therapy, they interact with the biomolecules and disturb the regular function of cells by aggregating inside the cells, forming a protein corona, and the NP-containing formulation turns ineffective in controlling cancer cell growth. The adverse interactions between NPs and biological entities can lead to life-threatening toxicities. Hence, there is indeed the potential for some NPs to be carcinogenic. It is a well-established fact that inhalation of NP results in overproduction of ROS, leading to reduction in antioxidant defense system and causing oxidative stress. This results in triggering inflammatory hormones and inflammation, which damages or deletes single-stranded as well as double-stranded DNA, inducing mutation and generating neoplasms and carcinoma in lungs. A survey and assessment done by Pope *et al.* [59] confirmed that the long-term exposure to NP leads to lung cancer and cardiopulmonary mortality in many metropolitan areas. Nanofibers and nanotubes that have high aspect ratio (that are analogous to asbestos, a known carcinogenic material) readily generate inflammation oxidants and mitogens that can lead to secondary carcinogenesis. Donaldson *et al.* [12] have suggested that reducing the length or aspect ratio may reduce the pathogenicity of nanorods, nanotubes, and nanofibers. However, low toxicity and low solubility NP, which are biologically not very active, are unlikely to pose a substantial cancer risk. Therefore, it is very important to check every NP as a separate issue pertaining to their carcinogenic potential; by assaying the hazards of NPs to the lung.

4.5 Lung Bioassay Methodologies for Assessing Hazards after Exposures to Nanoscale or Fine Particulates

We will now discuss lung bioassay methodology; or measurement of the concentration or potency of NPs by its effect on lung tissues; so as to assess the hazards of the NPs. The mechanism of exposure to NPs—inhalation, deposition, clearance, and the lung burden of NPs—are the important considerations for the bioassay of lung toxicity. There are different types of standard bioassay methods available, i.e., direct assays, indirect assays based upon quantitative responses; and indirect assays based upon quantal responses ("all or none").

Nanotoxicity testing is based on (i) dose response of a particular NP to which a person is exposed and (ii) exposure duration [51]. For studying toxicity of NP *in vitro*, assay methods ranging from endotoxin and lactate dehydrogenase (LDH) signaling to apoptosis and oxidative stress detection are used as they supply valuable techniques for exposing biomarkers of nanoparticle-induced cellular damage. For determining the toxicity of inhaled NPs, the solubility and chemical reactivity in the context of toxic actions are taken into consideration.

According to Warheit (2010) for conducting a pulmonary bioassay there is a need to have (i) robust characterization of the test material; (ii) a realistic dose-response paradigms utilizing relevant dose level settings that could be encountered in occupational or consumer environments; (iii) time course studies to ascertain the sustainability of any measured responses; and (iv) the inclusion of reference materials to better interpret the experimental findings that are derived from a study. Many methodologies and hypotheses have evolved over the years, which have offered an assessment of the health risk of inhaled NPs [2, 11, 17, 71, 87]).

These hypotheses for assaying adverse health effects of NP are based on: (i) Characteristics of NP, e.g., large surface area for interactions with cells and tissues, interaction with biomolecules, increased induction of oxidative stress and increased level of radical species compared to larger particles, induction of oxidative stress by lipid peroxidation and induction of cellular DNA damage. The NPs may differ in reactivity and solubility and may interact with all kinds of endogenous proteins, lipids, polysaccharides, and cells. (ii) Distribution of NPs as assessed by deposition characteristics dependent on size, uptake by cells of respiratory epithelium; nanotoxicity is evaluated by using epithelial cell lines of lung, skin, or gastrointestinal tract, as these cells present characteristics of real barriers against harmful agents and are therefore the first to suffer the influence of these compounds [65]. It also includes increased access to interstitial spaces and access to systemic circulation.

4.6 Strengths and Weaknesses of Risk Assessment

Previously, in Chapter 3 we discussed the use of control banding (CB) strategies for risk assessment of nanoparticles, because it offers simplified solutions for controlling worker exposures to constituents often found in the workplace, it minimizes or eliminates the need for complex quantitative assessments of worker exposure for comparison against occupational exposure standards and instead it provides a specific risk band based on a substance's hazard that is based on safety data sheets (SDSs).

The traditional risk assessment is done using cytotoxic studies, exposure measurement and computational methods; in four steps:

(i) Assessment of hazards;
(ii) Estimation of point of departure using a statistical model. Here, the "dose" is the airborne exposure concentration, resulting in the estimation of a benchmark concentration (maximum likelihood estimate) and a lower 95% confidence limit (BMCL) estimate;
(iii) Assessment of response to doses by estimating rat lung dose;
(iv) Exposure assessment by extrapolating the rat lung dose to a human-equivalent dose;
(v) Risk characterization.

Some of the major weaknesses in test guidelines and risk management are:

(i) Use of toxicological data from experimental studies is done in animals because human dose-dependent data is not available;
(ii) Absence of collaboration among researchers, industrial toxicologists, risk assessors, which is needed to obtain data to develop an improved basis for assessing the risk of exposure to nanomaterials;
(iii) Various components for risk assessment for different NPs, such as nanotubes, nanofibers, nanorods, spherical, cube, triangle, and hexagonal shaped, nanoclusters and different sizes of NPs. For example, risk assessment of inhaled CNTs are based on rat studies of pulmonary inflammation and fibrosis. The hazardous effect of SWCNT and MWCNT has been studies by many.
(iv) A gap exists between scientific knowledge and uncertainty towards environment, health, and safety risks due to the ever-increasing pace of commercialization of first generation nanoproducts and production. However, it must be said that efforts made in researching the health, safety and environmental concerns and risks of NPs have been stepped up in recent years. But our understanding of the interaction between NPs and living beings at the cellular level has not kept pace with the development of nanotechnology.

4.7 Concluding Remarks

Inhalation is one of the most dangerous routes of entry of NPs in the body, as the respiratory system is directly linked to the blood vascular system that aids in spreading the entry of NP to almost all the organs of the human body, including the central nervous system which governs all body functions. Our knowledge of the impact of NPs on human health and related safety issues of NPs are at the moment just above infancy stage. A lot of work and regulatory actions are needed regarding the inhalation toxicology database on nanomaterials, to draw far-ranging conclusions. It has been categorically shown that the same mass of many nano-size particles, such as carbon black, TiO_2, latex, etc., is 10-fold more inflammatory than that of their corresponding bulk or fine particles. In conclusion, I would like to say that emphasis should be placed on environmental protection against potentially toxic elements right from the stage of production, and also on dose-dependent inhalation aspects.

Bibliography

1. Asadi, F., Sadeghzadeh, M., Jalilvand, A., Nedaei, K., Asadi, Y., Heidari, A., Effect of molybdenum trioxide nanoparticles on ovary function in female rats. *J. Adv. Med. Biomed. Res.*, 27, 48–53, 2019, doi: 10.30699/jambs.27.121.48.
2. Bakand, S. and Hayes, A., Toxicological Considerations, Toxicity Assessment, and Risk Management of Inhaled Nanoparticles. *Int. J. Mol. Sci.*, 17, 6, 929, 2016, https://doi.org/10.3390/ijms17060929.
3. Banerjee, T. and Christian, R.A., A review on nanoparticle dispersion from vehicular exhaust: assessment of Indian urban environment. *Atmos. Pollut. Res.*, 9, 342–357, 2018, doi: 10.1016/j.apr.2017.10.009.
4. Barenholz, Y.C., Doxil®—the first FDA- approved nano-drug: lessons learned. *J. Control. Release*, 160, 117–134, 2012.
5. Barnes, P.J., Neurogenic inflammation in the airways. *Respir. Physiol.*, 125, 145–154, 2001.
6. Bisht, S. and Faiq, M., Tolahunase, M. *et al.*, Oxidative stress and male infertility. *Nat. Rev. Urol.*, 14, 470–485, 2017, https://doi.org/10.1038/nrurol.2017.69.
7. Buzea, C., Pacheco, I.I., Robbie, K., Nanomaterials and nanoparticles: sources and toxicity. *Biointerphases*, 2, MR17–MR71, 2007, doi: 10.1116/1.2815690.
8. Choi, H.S., *et al.*, Design considerations for tumour targeted nanoparticles. *Nat. Nanotechnol.*, 5, 42, 2010.

9. Clift, M.J., Gehr, P., Rothen-Rutishauser, B., Nanotoxicology: a perspective and discussion of whether or not *in vitro* testing is a valid alternative. *Arch. Toxicol.*, 85, 723–731, 2011, doi: 10.1007/s00204-010-0560-6.
10. Cruz, G.G.D.L., Rodríguez-Fragoso, P., Reyes-Esparza, J., Rodríguez, A., The Safety Profile of Nanoscale Particles and Materials - From Biomedical to Environmental Applications. *InTech*, doi:10.5772/intechopen.69386.
11. Da Silva, E., *The lung surfactant bioassay for hazard identification of airborne compounds: Principles, procedures and potential regulatory use*, Technical University of Denmark, 2021.
12. Donaldson, K., Murphy, F., Schinwald, A., Duffin, R., Poland, C.A., Identifying the pulmonary hazard of high aspect ratio nanoparticles to enable their safety-by-design. *Nanomedicine. (Lond)*, 6, 143–56, 2011.
13. Elsaesser, A. and Howard, C.V., Toxicology of nanoparticles. *Adv. Drug Deliv. Rev.*, 64, 129–137, 2012, doi: 10.1016/j.addr.2011.09.001.
14. Ermolin, M.S., Fedotov, P.S., Malik, N.A., Karandashev, V.K., Nanoparticles of volcanic ash as a carrier for toxic elements on the global scale. *Chemosphere*, 200, 16–22, 2018, doi: 10.1016/j.chemosphere.2018.02.089.
15. Feng, X., Chen, A., Zhang, Y., Wang, J., Shao, L., Wei, L., Central nervous system toxicity of metallic nanoparticles. *Int. J. Nanomed.*, 10, 4321, 2015, doi: 10.2147/IJN.S78308.
16. Ferreira, A.J., Cemlyn-Jones, J., Cordeiro, C.R., Nanoparticles, nanotechnology and pulmonary nanotoxicology. *Rev. Portuguesa Pneumol.*, 19, 28–37, 2013, doi: 10.1016/j.rppnen.2013.01.004.
17. Fröhlich, E. and Salar-Behzadi, S., Toxicological assessment of inhaled nanoparticles: role of *in vivo*, ex vivo, *in vitro*, and in silico studies. *Int. J. Mol. Sci.*, 15, 3, 4795–822, Mar 182014, doi: 10.3390/ijms15034795. PMID: 24646916; PMCID: PMC3975425.
18. Geiser, M., Update on macrophage clearance of inhaled micro and nanoparticles. *J. Aerosol Med. Pulm. Drug Deliv.*, 23, 207–217, 2010, doi: 10.1089/jamp.2009.0797.
19. Georgantzopoulou, A., Serchi, T., Cambier, S., Leclercq, C.C., Renaut, J., Shao, J., *et al.*, Effects of silver nanoparticles and ions on a co-culture model for the gastrointestinal epithelium. *Part. Fibre Toxicol.*, 13, 1–17, 2015, doi: 10.1186/s12989-016-0117-9.
20. Ghio, A.J., Kim, C., Devlin, R.B., Concentrated ambient air particles induce mild pulmonary inflammation in healthy human volunteers. *Am. J. Respir. Crit. Care Med.*, 162, 981– 988, 2000.
21. Gonzalez, C., Rosas-Hernandez, H., Ramirez-Lee, M.A., Salazar-García, S., Ali, S.F., Role of silver nanoparticles (AgNPs) on the cardiovascular system. *Arch. Toxicol.*, 90, 493–511, 2016, doi: 10.1007/s00204-014-1447-8.
22. Haase, A., Rott, S., Mantion, A., Graf, P., Plendl, J., Thünemann, A.F., *et al.*, Effects of silver nanoparticles on primary mixed neural cell cultures: uptake, oxidative stress and acute calciumresponses. *Toxicol. Sci.*, 126, 457–468, 2012, doi: 10.1093/toxsci/kfs003.

23. Chin-Lin, H., Hsiao, I.-L., Lin, H.-C., Wang, C.-F., Huang, Y.-J., Chuang, C.-Y., Silver nanoparticles affect on gene expression of inflammatory and neurodegenerative responses in mouse brain neural cells. *Environ. Res.*, 136, 253–263, 2015, ISSN 0013-9351, https://doi.org/10.1016/j.envres.2014.11.006.
24. Iavicoli, I., Fontana, L., Leso, V., Bergamaschi, A., The effects of nanomaterials as endocrine disruptors. *Int. J. Mol. Sci.*, 14, 16732–16801, 2013, doi: 10.3390/ijms140816732.
25. Ibald-Mulli, A., Stieber, J., Wichmann, H.-E., Koenig, W., Peters, A., Effects of air pollution on blood pressure: A population based approach. *Am. J. Public Health*, 91, 571–577, 2001.
26. Iversen, T.G., Skotland, T., Sandvig, K., Endocytosis and intracellular transport of nanoparticles: present knowledge and need for future studies. *Nano Today*, 6, 176–185, 2011, doi: 10.1016/j.nantod.2011.02.003.
27. Bonner, J.C., Nanoparticles as a potential cause of pleural and interstitial lung disease. *Proc. Am. Thorac. Soc.*, 7, 138–141, 2010, DOI: 10.1513/pats.200907-061RM.
28. Jaques Peter, A. and Kim Chong, S., Measurement of total lung deposition of inhaled ultrafine particles in healthy men and women. *Inhalation Toxicol.*, 12, 715–731, 2000, 0895±8378/00 $12.00 + .00.
29. Jiang, Z., Shan, K., Song, J., Liu, J., Rajendran, S., Pugazhendhi, A., *et al.*, Toxic effects of magnetic nanoparticles on normal cells and organs. *Life Sci.*, 220, 156–161, 2019, doi: 10.1016/j.lfs.2019.01.05.
30. Kim, J., Heo, Y.J., Shin, S., Hemocompatibility evaluation of silica nanomaterials using hemorheological measurements. *Clin. Hemorheol. Microcirc.*, 62, 99–107, 2016, doi: DOI: 10.3233/CH-151953.
31. Kreyling, W., Semmler, M., Erbe, F., Mayer, P., Takenaka, S., Schulz, H., *et al.*, Translocation of ultrafine insoluble iridium particles from lung epithelium to extrapulmonary organs is size- dependent but very low. *J. Toxicol. Environ. Health*, 65, 20, 1513–1530, 2002, doi: 10.1080/00984100290071649.
32. Kuhn, T., Biswas, S., Sioutas, C., Diurnal and seasonal characteristics of particle volatility and chemical composition in the vicinity of a light-duty vehicle freeway. *Atmos. Environ.*, 39, 7154–7166, 2005, doi: 10.1016/j.atmosenv.2005.08.025.
33. Laloy, J., Minet, V., Alpan, L., Mullier, F., Beken, S., Toussaint, O., Lucas, S., Dogné, J., Impact of silver nanoparticles on haemolysis, platelet function and coagulation. *Nanobiomedicine*, 4, 1–9, 2014, DOI: 10.5772/59346.
34. Leso, V., Fontana, L., Marinaccio, A., Leopold, K., Fanali, C., Lucchetti, D., *et al.*, Palladium nanoparticle effects on endocrine reproductive system of female rats. *Hum. Exp. Toxicol.*, 37, 1069–1079, 2018, doi: 10.1177/0960327118756722.
35. Li, C., Taneda, S., Taya, K., Watanabe, G., Li, X., Fujitani, Y., *et al.*, Effects of inhaled nanoparticle-rich diesel exhaust on regulation of testicular function in adult male rats. *Inhal. Toxicol.*, 21, 803–811, 2009, doi: 10.1080/089583708025 24381.

36. Nian, L., Li, Y., Liu, L., Liu, X., Yin, Y., Qu, G., Shi, J., Song, M., He, B., Hu, L., Jiang, G., Administration of silver nasal spray leads to nanoparticle accumulation in rat brain tissues. *Environ. Sci. Technol.*, 56, 1, 403–413, 2022, https://doi.org/10.1021/acs.est.1c02532.
37. Liu, J., *et al.*, Passive tumor targeting of renal- clearable luminescent gold nanoparticles: long tumor retention and fast normal tissue clearance. *J. Am. Chem. Soc.*, 135, 4978–4981, 2013.
38. Lozano-Fernández, T., Ballester-Antxordoki, L., Pérez-Temprano, N., Rojas, E., Sanz, D., Iglesias- Gaspar, M., Moya, S., González-Fernández, A., Rey, M., Potential impact of metal oxide nanoparticles on the immune system: The role of integrins, L-selectin and the chemokine receptor CXCR4. *Nanomedicine*, 10, 1301–1310, 2014, DOI: 10.1016/j.nano.2014.03.007.
39. Lu, X., Liu, Y., Kong, X., Lobie, P.E., Chen, C., Zhu, T., Nanotoxicity: a growing need for study in the endocrine system. *Small*, 9, 1654–1671, 2013, doi: 10.1002/smll.201201517.
40. Magdolenova, Z., Collins, A., Kumar, A., Dhawan, A., Stone, V., Dusinska, M., Mechanisms of genotoxicity. A review of *in vitro* and *in vivo* studies with engineered nanoparticles. *Nanotoxicology*, 8, 233–278, 2014, doi: 10.3109/17435390.2013.773464.
41. Medina, C., Santos-Martinez, M.J., Radomski, A., Corrigan, O.I., Radomski, M.W., Nanoparticles: pharmacological and toxicological significance. *Br. J. Pharmacol.*, 150, 552–558, 2007, doi: 10.1038/sj.bjp.0707130.
42. Mills, N.L., Miller, M.R., Lucking, A.J., Beveridge, J., Flint, L., Boere, A.J., Fokkens, P.H., Boon, N.A., Sandstrom, T., Blomberg, A., Duffin, R., Donaldson, K., Hadoke, P.W., Cassee, F.R., Newby, D.E., Combustion-derived nanoparticulate induces the adverse vascular effects of diesel exhaust inhalation. *Eur. Heart J.*, 32, 21, 2660–71, 2011 Nov, doi: 10.1093/eurheartj/ehr195.
43. Min, Y., Caster, J.M., Eblan, M.J., Wang, A.Z., Clinical translation of nanomedicine. *Chem. Rev.*, 115, 11147–11190, 2015.
44. Nemmar A., P.H.M., Vanquickenborne, B., Dinsdale, D., Thomeer, M., Hoylaerts, M.F., Vanbilloen, H., Mortelmans, L., Nemery, B., Passage of inhaled particles into the blood circulation in humans. *Circulation*, 105, 411–414, 2002.
45. Nho, R., Pathological effects of nano-sized particles on the respiratory system. *Nanomed. Nanotechnol. Biol. Med.*, 29, 102242, 2020, doi: 10.1016/j.nano.2020.102242.
46. O'loughlin, J., *et al.*, Safety, tolerability, and pharmacokinetics of SPL7013 gel (VivaGel): a dose ranging, phase I study. *Sex. Transm. Dis.*, 37, 100–104, 2010.
47. Oberdörster, G., Sharp, Z., Atudorei, V., Elder, A., Gelein, R., Kreyling, W., Cox, C., Translocation of Inhaled Ultrafine Particles to the Brain. *Inhalation Toxicol.*, 16, 437–445, 2004, DOI: 10.1080/08958370490439597.

48. Oberdörster, G., Oberdörster, E., Oberdörster, J., Invited review: Nanotechnology: An emerging discipline evolving from studies of ultrafine particles. *Environ. Health Perspect.*, 113, 7, 823–839, 2005a.
49. Oberdörster G., J., Ferin & B.E. Lehnert, Correlation between particle size, *in vivo* particle persistence, and lung injury. *Environ. Health Perspect.*, 102, Suppl. 5, 173–179, 1994.
50. Oberdörster, G., Sharp, Z., Atudorei, V., Elder, A., Gelein, R., Kreyling, W., Cox, C., Translocation of inhaled ultrafine particles to the brain. *Inhal. Toxicol.*, 16, 437–445, 2005b.
51. Oberdörster, G., Pulmonary effects of inhaled ultrafine particles. *Int. Arch. Occup. Environ. Health*, 74, 1–8, 2000, https://doi.org/10.1007/s004200000185.
52. Padmanabhan, J. and Kyriakides, T.R., Nanomaterials, inflammation, and tissue engineering. Wiley Interdisciplinary Review. *Nanomed. Nanobiotechnol.*, 7, 355–370, 2015, doi: 10.1002/wnan.1320.
53. Pakkanen, T.A., Kerminen, V.M., Korhonen, C.H., Hillamo, R.E., Aarnio, P., Koskentalo, T., *et al.*, Urban and rural ultrafine (PM0. 1) particles in the Helsinki area. *Atmos. Environ.*, 35, 4593–4607, 2001, doi: 10.1016/S1352-2310(01)00167-4.
54. Pekkanen J., Brunner, E.J., Anderson, H.R., Tiittanen, P., Atkinson, R.W., Daily concentrations of air pollution and plasma fibrinogen in London. *Occup. Environ. Med.*, 57, 818–822, 2000.
55. Peng, F., Setyawati, M.I., Tee, J.K., *et al.*, Nanoparticles promote *in vivo* breast cancer cell intravasation and extravasation by inducing endothelial leakiness. *Nat. Nanotechnol.*, 14, 279–286, 2019, https://doi.org/10.1038/s41565-018-0356-z.
56. Peters, A., Frohlich, M., Doring, A., Immervoll, T., Wichmann, H.-E., Hutchinson, W.L., Pepys, M.B., Koenig, W., Particulate air pollution is associated with an acute phase response in men; results from the MONICA-Augsburg Study. *Eur. Heart J.*, 22, 1198–1204, 2001.
57. Phillips, E., *et al.*, Clinical translation of an ultrasmall inorganic optical- PET imaging nanoparticle probe. *Sci. Transl. Med.*, 6, 260ra149, 2014.
58. Pope, C.A., Epidemiology of fine particulate air pollution and human health: biologic mechanisms and who's at risk? *Environ. Health Perspect.*, 108, 713–723, 2000.
59. Arden, P., II, Burnett, R.T., Turner, M.C., Cohen, A., Krewski, D., Jerrett, M., Gapstur, S.M., Thun, M.J., Lung cancer and cardiovascular disease mortality associated with ambient air pollution and cigarette smoke: Shape of the exposure–response relationships. *Environ. Health Perspect.*, 119, 11, 2011, https://doi.org/10.1289/ehp.1103639.
60. Pujalté, I., Passagne, I., Brouillaud, B., Tréguer, M., Durand, E., Ohayon-Courtès, C., L'Azou, B., Cytotoxicity and oxidative stress induced by different metallic nanoparticles on human kidney cells. *Part. Fibre Toxicol.*, 8, 10, 2011, doi: 10.1186/1743-8977-8-10.

61. Qiao, H., Liu, W., Gu, H., Wang, D., Wang, Y., The transport and deposition of nanoparticles in respiratory system by inhalation. *J. Nanomater.*, 2015, 394507, 2015, doi: 10.1155/2015/394507.
62. Reddy, L.H., Arias, J.L., Nicolas, J., Couvreur, P., Magnetic nanoparticles: design and characterization, toxicity and biocompatibility, pharmaceutical and biomedical applications. *Chem. Rev.*, 112, 5818–5878, 2012.
63. Renwick, L.C., Donaldson, K., Clouter, A., Impairment of alveolar macrophage phagocytosis by ultrafine particles. *Toxicol. Appl. Pharmacol.*, 172, 119–127, 2001.
64. Riipinen, I., Yli-Juuti, T., Pierce, J.R., Petäjä, T., Worsnop, D.R., Kulmala, M., *et al.*, The contribution of organics to atmospheric nanoparticle growth. *Nat. Geosci.*, 5, 453–458, 2012, doi: 10.1038/ngeo1499.
65. Rothen-Rutishauser, B., Clift, M.J.D., Jud, C., Fink, A., Wick, P., Human epithelial cells *in vitro* - Are they an advantageous tool to help understand the nanomaterial-biological barrier interaction? *EURO-NanoTox-Letters*, 4, 1, 1–19, 2015.
66. Hayes, A. and Dechsakulthorn, F., Nanoparticles: a review of particle toxicology following inhalation exposure. *Inhal. Toxicol.*, 24, 125–135, 2012, doi: 10.3109/08958378.2010.642021.
67. Salvi, S., Blomberg, A., Rudell, B., Kelly, F., Sandstrom, T., Holgate, S.T., Frew, A., Acute inflammatory responses in the airways and peripheral blood after short-term exposure to diesel exhaust in healthy human volunteers. *Am. J. Respir. Crit. Care Med.*, 159, 702–709, 1999.
68. Salvi, S., Nordenhall, C., Blomberg, A., Rudell, B., Pourazer, J., Kelly, F.J., Wilson, S., Sandstrom, T., Holgate, S.T., Frew, A., Acute exposure to diesel exhaust increases IL-8 and GRO-a production in healthy human airways. *Am. J. Respir.Crit. Care Med.*, 161, 550–557, 2000.
69. Santonastaso, M., Mottola, F., Colacurci, N., Iovine, C., Pacifico, S., Cammarota, M., *et al.*, *In vitro* genotoxic effects of titanium dioxide nanoparticles (n-TiO2) in human sperm cells. *Mol. Reprod. Dev*, 86, 1369–1377, 2019, doi: 10.1002/mrd.23134.
70. Sardar, S.B., Fine, P.M., Mayo, P.R., Sioutas, C., Sizefractionated measurements of ambient ultrafine particle chemical composition in Los Angeles using the NanoMOUDI. *Environ. Sci. Technol.*, 39, 932–944, 2005, doi: 10.1021/es049478j.
71. Savage, D.T., Hilt, J.Z., Dziubla, T.D., *In vitro* methods for assessing nanoparticle toxicity. *Methods Mol. Biol.*, 1894, 1–29, 2019, doi: 10.1007/978-1-4939-8916-4_1. PMID: 30547452; PMCID: PMC6984383.
72. Schwartz, J., Air pollution and blood markers of cartdiovascular risk. *Environ. Health Perspect.*, 109, Suppl. 3, 405–409, 2001.
73. Seaton, A., Soutar, A., Crawford, V., Elton, R., McNerlan, S., Cherrie, J., Watt, M., Agius, R., Stout, R., Particulate air pollution and the blood. *Thorax*, 54, 1027–1032, 1999.

74. Sela, H., *et al.*, Spontaneous penetration of gold nanoparticles through the blood brain barrier (BBB). *J. Nanobiotechnol.*, 13, 71, 2015.
75. Sharma, V., Singh, P., Pandey, A.K., Dhawan, A., Induction of oxidative stress, DNA damage and apoptosis in mouse liver after sub-acute oral exposure to zinc oxide nanoparticles. *Mutat. Res. Genet. Toxicol. Environ. Mutagen.*, 745, 84–91, 2012, doi: 10.1016/j.mrgentox.2011.12.009.
76. Sonwani, S., Madaan, S., Arora, J., Suryanarayan, S., Rangra, D., Mongia, N., Vats T and Saxena, P., Inhalation Exposure to Atmospheric Nanoparticles and Its Associated Impacts on Human Health: A Review. *Front. Sustain. Cities*, 3, 690444, 2021, doi: 10.3389/frsc.2021.690444.
77. Sonwani, S. and Saxena, P., Water-insoluble carbonaceous components in rainwater over an urban background location in Northern India during pre-monsoon and monsoon seasons. *Environ. Sci. Pollut. Res.*, 1–16, 2021, doi: 10.1007/s11356-021-14132-w.
78. Strambeanu, N., Demetrovici, L., Dragos, D., Anthropogenic sources of nanoparticles, in: *Nanoparticles' Promises and Risks*, vol. 21–54, Springer, Cham, 2015.
79. Stueckle, T.A., Davidson, D.C., Derk, R., Kornberg, T.G., Schwegler-Berry, D., Pirela, S.V., *et al.*, Evaluation of tumorigenic potential of CeO2 and Fe2O3 engineered nanoparticles by a human cell *in vitro* screening model. *Nanoimpact*, 6, 39–54, 2017, doi: 10.1016/j.impact.2016.11.001.
80. Utembe, W., Tlotleng, N., Kamng'ona, A.W., A systematic review on the effects of nanomaterials on gut microbiota. *Curr. Res. Microb. Sci.*, 3, 100118, 2022. ISSN 2666- 5174, https://doi.org/10.1016/j.crmicr.2022.100118.
81. Valdiglesias, V., Costa, C., Kiliç, G., Costa, S., Pásaro, E., Laffon, B., *et al.*, Neuronal cytotoxicity and genotoxicity induced by zinc oxide nanoparticles. *Environ. Int.*, 55, 92–100, 2013, doi: 10.1016/j.envint.201. 3.02.013.
82. Venkatesan, B., Tumala, A., Subramanian, V., Vellaichamy, E., Data on synthesis and characterization of chitosan nanoparticles for *in vivo* delivery of siRNA-Npr3: Targeting NPR-C expression in the heart. *Data Brief*, 8, 441–7, 2016 Jun 3, doi: 10.1016/j.dib.2016.05.074. PMID: 27366782; PMCID: PMC4910299.
83. Vincent, R., Kumarathasan, P., Goegan, P., Bjarnason, S.G., Guenette, J., Berube, D., Adamson, I.Y., Desjardins, S., Burnett, R.T., Miller, F.J., Battistini, B., Inhalation Toxicology of Urban Ambient Particulate Matter: Acute Cardiovascular Effects in Rats. Research Report 104, Health Effects Institute, Boston, MA, 2001.
84. Walsh, M., Global trends in motor vehicle pollution control: a 2011 update. Part 1. *Silniki Spalinowe*, 50, 106–117, 2011, doi: 10.19206/CE-117109.
85. Wang, R., Song, B., Wu, J., Zhang, Y., Chen, A., Shao, L., Potential adverse effects of nanoparticles on the reproductive system. *Int. J. Nanomed.*, 13, 8487–8506, 2018.
86. Warheit, D.B., Laurence, B.R., Reed, K.L., Roach, D.H., Reynolds, G.A.M., Webb, T.R., Comparative Pulmonary Toxicity Assessment of Single-wall

Carbon Nanotubes in Rats. *Toxicol. Sci.*, 771, 117–125, January 2004, https://doi.org/10.1093/toxsci/kfg228.
87. Warheit, D.B., *et al.*, Nanoparticles health impact. *Toxicologist*, 72, 1, 298A, 2003.
88. Warheit, D.B., Assessing health risks of inhaled nanomaterials: development of pulmonary bioassay hazard studies. *Anal. Bioanal. Chem.*, 398, 607–612, 2010, https://doi.org/10.1007/s00216-010-4000-5.
89. Yang, Z., Liu, Z.W., Allaker, R.P., Reip, P., Oxford, J., Ahmad, Z., *et al.*, A review of nanoparticle functionality and toxicity on the central nervous system. *J. Roy. Soc. Interface*, 7, S411–S422, 2010, doi: 10.1098/rsif.2010.0158.focus.
90. Yu, X., Hong, F., Zhang, Y.Q., Bio-effect of nanoparticles in the cardiovascular system. *J. Biomed. Mater. Res. Part A*, 104, 2881–2897, 2016, doi: 10.1002/jbm.a.35804.
91. Zhou, H. and McClements, D.J., Recent Advances in the Gastrointestinal Fate of Organic and Inorganic Nanoparticles in Foods. *Nanomater. (Basel)*, 12, 7, 1099, 2022 Mar 27, doi: 10.3390/nano12071099. PMID: 35407216; PMCID: PMC9000219.
92. Zhou, C., *et al.*, Near-infrared emitting radioactive gold nanoparticles with molecular pharmacokinetics. *Angew. Chem. Int. Ed. Engl.*, 124, 10265–10269, 2012.
93. Zhu, X., Hondroulis, E., Liu, W., C.Z., Li, Biosensing approaches for rapid genotoxicity and cytotoxicity assays upon nanomaterial exposure. *Small*, 9, 1821–1830, 2913, doi: 10.1002/smll.201201593.

5

Dermal Entry Route of NPs into the Human Body: A Risk?

In thinking about nanotechnology today, what's most important is understanding where it leads, what nanotechnology will look like after we reach the assembler breakthrough.

K. Eric Drexler

Abstract

The skin, once thought to be a strong barrier, is now recognized as a vulnerable entry point for nanoparticles—especially from cosmetics and pharmaceuticals. This chapter examines how nanoparticles penetrate healthy and damaged skin, interact with biomolecules, and vary in absorption based on size, shape, surface properties, and environmental factors like pH and UV exposure. It highlights concerns around dermal nanotoxicity, emphasizes the need to assess risks tied to diverse nanoparticle types, and calls for improved evaluation of skin exposure hazards.

***Keywords*:** Dermal, skin-appendages, pathways, entry-route, stratum corneum, human-body, dermal-exposure

5.1 Introduction

As we know, nanotechnology is the fabrication and application of NPs and derived devices or material at the nanoscale (1–100 nm), with unique characteristics, which are different from those in their counterparts at bulk or macroscale. The airborne nanoparticles may be due to incidentally formed nanoparticles, sometimes also called ultrafine particles; or engineered nanoparticles, that are intentionally designed and produced. An interface between the environment containing NPs and the individual exposed to them over a specific time are of relevant concern as they may cause adverse health outcomes. Hence, there is a need to implement hazard/toxicity

Madhuri Sharon. Risk, Safety, and Ethical Concerns in Nanotechnology, (89–102) © 2025 Scrivener Publishing LLC

studies. In the previous chapter I have discussed the inhalation route of exposure to NPs; however, dermal and oral routes of entry of NPs are equally important. Therefore, this critical route of entry of NPs is discussed in this chapter to assess the health impacts. Moreover, workplace exposure assessments, methodologies and protocols should be considered to better ascertain the route and impact of NPs on dermal absorption and cutaneous toxicity. However, comparatively less data is available regarding the health risks of dermal exposures to nanomaterials, which may be due to dermal absorption and cutaneous toxicity. Hence, there is a need to implement the hazard/toxicity studies. Moreover, workplace exposure assessments, methodologies and protocols should be considered to better ascertain the route and impact of NPs on dermal absorption and cutaneous toxicity.

According to the International Organization for Standardization (ISO) [18], Nanoparticles (NPs) that are discussed here are *"nano-objects with all three external dimensions in the nanoscale"* where the nanoscale was defined as the size ranging from 1 nm to 100 nm. The three dimensions are external dimension, internal structure, and surface structure. Whereas, Nanomaterial is defined as a *"material with any external dimension in the nanoscale or having internal structure or surface structure in the nanoscale"* [17].

Though nanoparticles (NPs) have been found to penetrate through wounded and cut skin, Labouta and Schneider [21] had expressed uncertainty about skin being a significant route for entry of nanoparticles (NPs). The primary barrier to chemical absorption is the *stratum corneum* of the skin (Figure 5.1). Comparatively less data is available regarding the health risks of dermal exposures to nanomaterials (NMs), which may be due to dermal absorption and cutaneous toxicity. Other routes through which

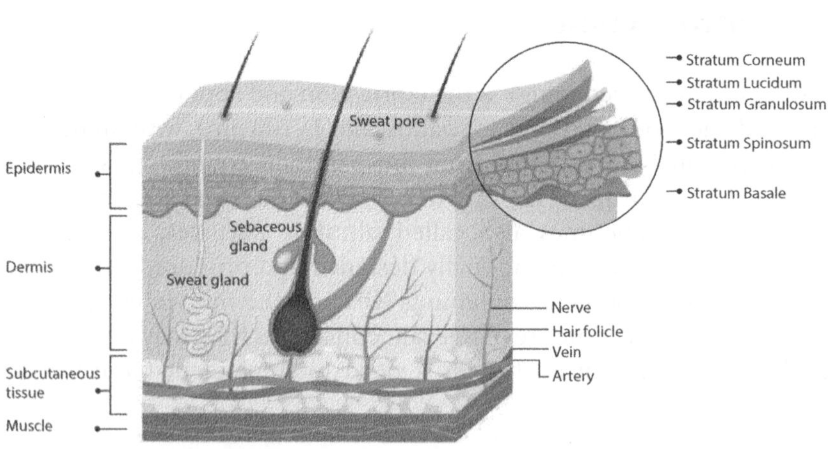

Figure 5.1 Diagram of human skin anatomy.

highly dispersible NPs get entry into the human body, such as ocular and hair follicle, can be considered as affecting dermal exposure through both airborne routes as well as its deposition on working surfaces. Another concern is overall weighting of the dermal toxicity parameters. Moreover, the potential of dermal penetration of NMs through intact skin, needs consideration along with the factors that encompass cutaneous toxicity. Moreover, skin has many layers (Figure 5.1), and whether all types of NPs can penetrate through all of them needs thorough evaluation because the interaction of nanoparticles with skin as barrier as well as sink function to artificial NP compounds is an intriguing field. The potential of NPs in the medical and pharmaceutical fields cannot be denied, but on the other hand, health hazards are also to be identified and assessed. Dermally absorbed NPs have been found to have local and systemic toxicity.

5.2 Dermal Entry Route of Nanoparticles

Skin is a protective organ composed of the epidermis, hypodermis, and endodermis, along with derivatives such as sweat glands, oil glands, and the hair embedded in it (Figure 5.1), which offers a barrier to unwanted entities, bacteria, fungi, chemicals and even temperature to some extent. Whether NPs can penetrate the skin has become an important concern, because many NP-based products that are applied to the skin are now on the market. Blood capillaries that run close to skin may assist in translocating NPs to different organs once they are taken up through the skin. But there are reports suggesting that some inorganic NPs, such as titanium dioxide present in sunscreen cream, can penetrate through the stratum corneum and the dermis [22, 26, 43, 50].

For entry through the skin, mostly pharmaceutical and cosmetic NP-containing products are being considered. But apart from cosmetics, there are many commercially available products, such as sunscreens, stain-resistant clothing, sports equipment, automobile catalytic converters, dental bonding, cleaning products, and dressings for specific wound care, that come in close contact with skin. Moreover, there are some formulations of NPs that contain dimethyl sulfoxide, which alters the permeability of the skin by removing much of the lipid matrix of the stratum corneum, leaving holes and shunts. Suitable organic NPs that are nontoxic for transdermal uptake are micelles, dendrimers, liposomes, and other polymers, especially because they are in the form of nanospheres or nano-capsules [20, 35, 47]. The NPs prepared from or extracted from biological systems have also been found suitable and nontoxic [19]. The most commonly used inorganic (metal or metal oxide) NPs are aluminum, titanium, gold, iron, silver, zinc, copper, and their oxides. Though as such they are nontoxic, it has been found that if there are cuts, wounds, or lacerations, they facilitate entry of foreign bodies such as NPs [42], resulting in adverse effects. Penetration of NPs through the skin is influenced by the size, shape, composition, surface coating and geometry. A recent review about skin absorption of inorganic nanoparticles and their toxicity [37] has summarized the cutaneous absorption of different types of NPs using *in-vivo* and *in-vitro* methodologies on the skin of different animals. Based on *in-vitro* studies, it was concluded that the NPs from sun-screen creams can barely pass the stratum corneum of the skin; and exposure can be prevented with common gloves and clothes. However, I think that there is a need to extend these studies to more NPs to be incorporated in the future skin absorption evaluation of their health risk and environmental impact. Rather than using animals for evaluation, a mimic membrane should be developed to evaluate the percutaneous NPs absorption. Skin is selectively permeable to quantum dots and some nanoparticles [27].

5.2.1 Pathways for Skin Absorption of Nanoparticles

Depending on the physicochemical properties of NPs, absorption of NPs through the skin can be envisaged using two pathways: (i) through the stratum corneum by intercellular or transcellular mechanism, and (ii) through the skin appendages by the trans-appendageal route (Figure 5.2).

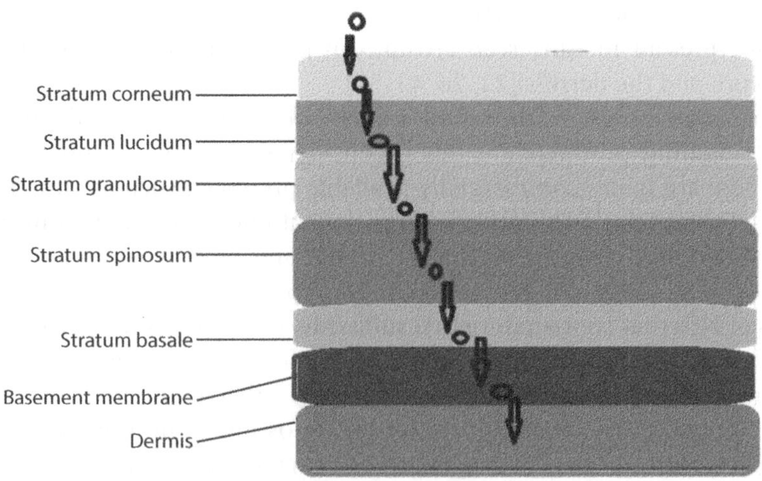

Figure 5.2 Diagram showing the pathways for nanoparticle exposure and translocation through human skin.

5.2.1.1 *Stratum Corneum (SC)*

The stratum corneum (SC) is assumed to be the main barrier for absorption into the underlying tissues of the skin. The SC is composed of rigid corneocytes, dead cells packed with keratinous filaments in a matrix of connecting proteins and confined by the cornified envelope, separated by the intercellular space filled with lipid compounds and aqueous films to form lamellar lipid bilayer; whereas the underlying living tissues are in an aqueous environment. Hence, lipophilic molecules can distribute more easily in the SC and penetration is facilitated. Nanoparticles < 40 nm have been found to easily penetrate the SC through the lipidic intercellular route. To move along the lipid or aqueous phase absorbed NPs need available intercellular space, that is restricted and therefore the size of penetrating NPs influences the absorption behavior. Moghini *et al.* (1999) has suggested that only substances with partition coefficients (logP) between 1 and 3 are well suited for skin absorption.

5.2.1.2 *Skin Appendages*

The skin appendages, i.e., epidermal and dermal-derived components like hair follicles, nails, sweat glands, sebaceous glands and/or pores; may be other routes through which NPs may enter at a faster rate due to altered barrier morphology, and can reach deeper into the skin. The process of invasion through the skin appendages is not expected to be by absorption. The processes responsible for NP penetration is not yet fully understood. However, let us look at the interaction of absorbed NPs with the biomolecules.

5.2.1.3 *Interaction and Impact of Dermally Absorbed Nanoparticles with Biomolecules*

Once the NPs are absorbed, they interact with proteins, which may alter the behavior of protein [9], which in turn may induce adverse genotoxic effects either directly or indirectly (mediated inflammatory processes) causing oxidative stress due to the formation of active oxygen species. The adverse effects are based on the properties of NP formulations and their interaction mechanism; whether they have entered through the intercellular route, i.e., the lipid channels between the corneocytes to deeper skin layers or through the appendage route [1, 2, 23, 52]. The impact of NPs used in dermal cream or ointment depends on their chemical composition, which can be either polymeric, metallic or nano-forms of carbon. For example:

(i) *Lipids or Solid Lipid NPs (SLN)-Containing Drugs:* When applied onto skin, these result in higher drug permeation due to enhanced drug absorption as well as due to the occlusive effect as a result of surface coverage [46], which causes complete or partial blockage of a blood vessel, reducing the flow of oxygen-rich blood to the heart, brain, etc. In their study, Schneider *et al.* [41] mentioned that there has been no intact penetration of the SLN reported so far.

(ii) *Liposomes:* These are a very common group of nanoparticulate carriers used in cosmetics or for therapeutic purposes. They are also a lipid component composed of a closed bilayer of phospholipids, i.e., hydrophobic compartment (lipid layer) and a hydrophilic compartment (inner volume of the liposome), which facilitates incorporation of a wide variety of drugs as well as offers the biocompatibility inherently connected with natural phospholipids. Liposomes are ≥50 nm in size, which is bigger than the skin openings; therefore, passive penetration is not possible. To penetrate the skin a driving force is needed [8], which suggests that a water activity gradient in the stratum corneum may be responsible for the motion of the liposomes along the pores of the intercellular space.

(iii) *Protein Particles:* These are not yet being used for dermal application; however, they are being commercially used for some oral application, through which skin uptake takes place.

(iv) *Metal and Metal Oxide Nanoparticles:* There are many metallic NPs that are used in cosmetics/skincare products such as gold, silver, beryllium, titanium dioxide, zinc oxide, maghemite (γ-Fe2O3) and iron core shell particles. Many cosmetic products containing gold nanoparticles (AuNP) are already on the market, e.g., Gold Skincare sets by Temu containing moisturizing serum with nicotinamide for firming skin and as anti-wrinkle cream. Protein (albumin, transferrin) functionalized AuNP to supplement skin protein by Luna Nanotech and many more. Gold is preferred due to ease of preparing AuNP of different sizes as well as its easy functionalization. Sonavane *et al.* [45] have shown that the smallest particles tend to aggregate once they enter into deeper skin layers, whereas the larger particles could reach up to the epidermis and dermis. Beryllium has been found in the stratum corneum [51], whereas, maghemite (γ-Fe_2O_3) and iron core shell particles (d $\leq$ 20 nm) penetrated through the top layer to the viable epidermis and hair follicle of human skin [3]. There are other metallic NPs, such as silver, that could easily penetrate skin. However, data on the adverse effects of these NPs are sparse. Follicular penetration also needs to be studied.

Metal oxides like titanium dioxide and zinc oxide are used in sunscreen lotion in a size range of 50–500 nm. According to SCCNP (Scientific Committee on Cosmetics and Non-Food Products), passive diffusion of these NPs does not occur and they remain on the skin surface or on the outer layer of the stratum corneum and do not penetrate into or through the living skin [23, 50].

(v) Though there are some activated carbon-containing skin care products on the market, the use of carbon nanotube in dermal product has not come on the market. There are limited data on carbon-based nanoparticles.

5.3 Health Hazards of Dermal Exposure to NPs

There are many concerns that are related to the dermal use of NPs, that are not clear from the health point of view, e.g., can agglomerates (~250 nm) penetrate or only the primary particles penetrate. It is essential to investigate so that the behavior of NP with skin can be predicted, because even slight differences in the treatment of the skin can result in altered penetration behavior. Studies have shown that spherical particles go down deeper in the epidermis and dermis whereas ellipsoidal particles are mainly found in the stratum corneum and epidermal layers [36]. Other factors that affect dermal absorption of NP are found to be the pKa of the penetrating molecule [4]; solubility and dissolved amount of the penetrating molecule [5]; partition coefficient of the penetrating molecule [30]; molecular weight of a penetrating agent, which should be less than 600 Da to significantly permeate skin [6]; potential for binding and metabolism [32] and diffusion coefficient (D) of the penetrating molecule [4].

The primary route of exposure for workers involved in manufacturing NPs, is mostly through inhalation and/or dermal contact. Dermal exposure is less likely to occur during the manufacturing, since NP manufacturing is done under controlled conditions and is performed in closed reaction chambers.

There is a chance of dermal exposure to NPs while using skin care products. The dermal penetration capacity of various NPs has led to cytotoxicity studies of various NPs. The uptake and hazardous impact of dermally absorbed NPs is related to the skin condition, such as, stratum corneum integrity, atopic-dermatitis, psoriasis, ichthyosis, inflammation, burns, infections, damage, trauma, extensive use of detergents, dry skin caused by prolonged exposure to non-humidified environments, etc.

Carbon Nanotube: Both single-wall carbon nanotube (SWCNT) and multi-wall carbon nanotube (MWCNT) were found to be toxic when exposed to the skin. It has been reported that SWCNT causes (i) Oxidative stress and cellular toxicity in human epidermal Keratinocytes at a concentration range of 0.6–0.24 mg mL^{-1} [44]; (ii) Dose and time-dependent inhibition of cell proliferation and decreased cell adhesive ability in kidney cells of human embryo at 0.8–200 μg mL^{-1} [11]; (iii) Degree of functionalization of water-soluble MWCNT was found to be another criteria that decides the response by human dermal fibroblast cells *in vitro.* Different functionalized groups were tried, i.e., SWNT-phenyl-SO(3)H, SWNT-phenyl-SO(3)Na, SWNT-phenyl-(COOH)(2) (one sample with carbon/-phenyl-(COOH)(2), and underivatized SWNT stabilized in 1% Pluronic F108. The results showed that with an increase in the degree of sidewall functionalization the SWCNT sample becomes less cytotoxic [38]. MWCNT has been found to cause a decrease in human epidermal keratinocytes at 0.1–0.4 mg mL^{-1} within 1–48 hours [28]. There is a need for careful documentation of the physical and chemical characteristics of CNTs used and their health and environmental impact.

Metal Oxide: Two commonly used metal oxides for dermal application have shown adverse health effects; for example, (i) at 3.6 μg mL^{-1} Anatase TiO_2 NPs killed human fibroblast cell [39], and (ii) at 25 μg mL^{-1} SiO_2 NPs significantly inhibit replication and transcription in human epithelial cells [9].

5.4 Risk Handling and Evaluation of Dermal Exposure Hazards

Currently we do not have a perfect risk evaluation method so that the regulation of nanomaterials can be established in a foolproof way to foresee the risk to health. However, many approaches for nanomaterial regulation are being developed in some countries; for example, the Environmental Protection Agency (EPA) in the USA in collaboration with the Food and

Drug Administration (FDA) [16], and ECHA working closely with the European Union Member States [14].

Usually, studies on dermal penetration of NPs is limited to consumer products such as sunscreen. But there is a need for precaution and safety studies on potential photooxidation products, the effect of skin condition on penetration, specific assessments of potential hazards, exposure, and likelihood of toxic effects. For risk assessment the first step should be to identify the hazard by knowing the particle size and structure of the NP, and its chemical composition, properties, and coating (encapsulation) or functionalizing moieties. This will help in exposure and toxicity assessment. The exposure assessment encompasses behaviors of NP, product uses, durability of product, route of entry and whether there is any receptor during the entire NP cycle, from synthesis to disposal. To assess the toxicity of NP, it is important to know its uptake, its distribution in the tissue dosimetry, whether it takes part in metabolism and what its reactivity is, its biological interaction with the metabolites, and finally how much of it gets excreted from the system and the amount of time it takes. This analysis will help in risk characterization such as likelihood of effect, nature of effect, and effectiveness of control [40].

During the synthesis and manufacture of products containing NPs, there is potential for their release into the workplace environment, leading to occupational human exposure through inhalation. Finally, attrition and disposal are involved in ecological and environmental exposure. Many NPs, such as CNTs and metal oxides, tend to aggregate, which may reduce their ability to penetrate membrane or disperse in air. However, exposure varies as per specific product.

Some of the important risk assessment studies of NPs *in vivo* have shown that NP exposure in animals can be seen on the skin and analyzed *via* a skin biopsy [48].

Tak *et al.* have studied the importance of AgNP shape in dermal penetration [49]. Comparatively more studies have been done on *in-vitro* penetration through skin using epidermal keratinocyte HaCaT cells [10, 24]. The results have shown that (i) Physicochemical conditions like pH and light exposure enables dermal penetration of inorganic NPs like TiO_2 [7] as well as toxicity [12]. (ii) Size and surface charge of silica NP caused skin toxicity [33], and amorphous silica NPs induced consumptive coagulopathy [31]. (iii) Many NPs have adverse effects; for example, ZnO NPs induce apoptosis [25]; and citrate-capped gold adversely affects human dermal fibroblast [34]. (iv) Structure and function of skin has also been found to be related to the toxicity [29].

5.5 Conclusion

Skin is a protective permeable barrier against any external agent, be it bacteria, fungus, virus, chemicals, light or UV light. It also regulates body temperature and the Langerhans cells of skin regulate the immunological response. Since 2006, skin has been facing many problems created by NPs of metals, metal oxides, CNTs, etc. The source of these NPs are mostly cosmetics, pharmaceutical material, sunscreen creams, antibacterial creams, etc. There are studies that declare these NPs to be safe and do not cross the skin barrier; whereas other studies have shown that iron NPs *via* hair follicles [13] could reach the basal and spinous layers [53]. Another aspect that has not been perfected is a safety profile evaluation of the hazardous impact of different types of NPs that vary in size, shape, structure, and functionalized units. These variations influence the entry of NPs through the skin; for example, NPs of 4 nm diameter could penetrate intact skin, whereas 45 nm NPs can only permeate damaged skin [15]. Both natural and artificial radiation causes a disaggregation of TiO_2 NPs; even a few minutes of irradiation leads to a decrease in TiO_2 aggregates from 280 nm to 230 nm. Photoinduced disaggregation of TiO_2 NPs to enable transdermal penetration was suggested by Bennet *et al.* [7], and the role of degradation in different pH conditions and light exposure was reported by De Matteis *et al.* [12].

With an increase in the business of NP applications in cosmetics and medicine, despite the unique properties of NPs, their dark side has to be evaluated. This is especially true in relation to the environment since their adverse effects to human health are toxic over time due to the perpetual stimulation of the immune system, which induces an inflammatory condition, especially in those working on nanomaterial production. *In-vitro* and *in-vivo* studies on NP penetration through the dermal layer, along with predictive computational toxicity methods, are providing data related to the adverse effects of NPs. Efforts are underway to modify the surface of NPs so as to make them less toxic and safer in order to produce suitable consumer products. This requires extensive standardized *in-vitro* procedures to establish the toxicology profile of NPs; amd the implementation of specific standard operating procedures (SOPs) for the testing of NPs based on good laboratory practices (GLP). There is a need to create flexible and reliable databases in which NPs can be classified based on results derived from toxicological studies for different types of NPs, their safe concentrations, the type of toxicity, and development of protective devices required to stop the entry into the body. Moreover, clear legislation on the use of nanomaterials has to be formulated.

References

1. Alvarez-Roman, R., Naik, A., Kalia, Y., Guy, R.H., Fessi, H., Skin penetration and distribution of polymeric nanoparticles. *J. Control Release*, 99, 53–62, 2004b.
2. Alvarez-Roman, R., Naik, A., Kalia, Y.N., Fessi, H., Guy, R.H., Visualization of skin penetration using confocal laser scanning microscopy. *Eur. J. Pharm. Biopharm.*, 58, 301–16, 2004a.
3. Baroli, B., Ennas, M.G., Loffredo, F., Isola, M., Pinna, R., Lopez-Quintela, M.A., Penetration of metallic nanoparticles in human full-thickness skin. Journal of Investigative. *Dermatology*, 127, 1701–12, 2007.
4. Baroli, B., *J. Pharm. Sci.*, 99, 21–50, 2010.
5. Barry, B.W., *Dermatological formulations*, pp. 48–94, Marcel Dekker, Inc., New York, 1983.
6. Barry, B.W., *Nat. Biotechnol.*, 22, 165–167, 2004.
7. Bennett, S.W., Zhou, D., Mielke, R., Keller, A.A., Photoinduced Disaggregation of TiO2 Nanoparticles Enables Transdermal Penetration. *PLoS One*, 2012, 7, e48719.
8. Cevc, G. and Gebauer, D., Hydration-driven transport of deformable lipid vesicles through fine pores and the skin barrier. *Biophys. J.*, 84, 1010–24, 2003.
9. Chen, M. and von Mikecz, A., Formation of nucleoplasmic protein aggregates impairs nuclear function in response to SiO2 nanoparticles. *Exp. Cell. Res.*, 305, 51–62, 2005.
10. Crosera, M., Prodi, A., Mauro, M., Pelin, M., Florio, C., Bellomo, F., Adami, G., Apostoli, P., De Palma, G., Bovenzi, M., *et al.*, Titanium Dioxide Nanoparticle Penetration into the Skin and Effects on HaCaT Cells. *Int. J. Environ. Res. Public Health*, 12, 9282–9297, 2015.
11. Cui, D., Tian, F., Ozkan, C., Wang, M., Gao, H., Effect of single wall carbon nanotubes on human HEK293 cells. *Toxicol. Lett.*, 155, 73–85, 2005.
12. De Matteis, V., Cascione, M., Brunetti, V., Toma, C.C., Rinaldi, R., Toxicity assessment of anatase and rutile titanium dioxide nanoparticles: The role of degradation in different pH conditions and light exposure. *Toxicol. Vitro*, 37, 201–210, 2016.
13. Donnelly, R.F., Singh, T.R.R., Morrow, D.I., Woolfson, A.D., *Microneedle-Mediated Transdermal and Intradermal Drug Delivery*, pp. 1–216, Wiley-Blackwell, Hoboken, NJ, USA, 2012.
14. European Commission Web Resource. Available online: http://ec.europa.eu/research/industrial_technologies/policy_en.html (accessed on 12 August 2016).
15. Filon, F.L., Mauro, M., Adami, G., Bovenzi, M., Crosera, M., Nanoparticles skin absorption: New aspects for a safety profile evaluation. *Regul. Toxicol. Pharmacol.*, 72, 310–322, 2015.

16. Hegde, K., Brar, S.K., Verma, M., Surampalli, R.Y., Current understandings of toxicity, risks and regulations of engineered nanoparticles with respect to environmental microorganisms. *Nanotechnol. Environ. Eng.*, 1, 5, 2016.
17. ISO 2010 - Nanotechnologies Vocabulary Part 1: Core Terms. ISO/TS 80004-1: 2010. 2010. Available online: https://www.iso.org/standard/51240.html (accessed on 12 October 2017).
18. ISO 208-Technical Specification: Nanotechnologies Terminology and Definitions for Nano-Objects Nanoparticle, Nanofibre and Nanoplate. ISO/TS 80004-2: 2008. 2008. Available online: https://www.iso.org/standard/44278.html.
19. Jutz, G. and Böker, A., Bionanoparticles as functional macromolecular building blocks–a new class of nanomaterials. *Polymer*, 52, 2, 211–232, 2011. https://doi.org/10.1016/j.polymer.2010.11.047.
20. Kumar, R. and Lal, S., Synthesis of organic nanoparticles and their applications in drug delivery and food nanotechnology: a review. *J. Nanomater. Mol. Nanotechnol.*, 3, 4, 2014. https://doi.org/10.4172/2324-8777.1000150.
21. Labouta Hagar, I. and Schneider, M., Interaction of inorganic nanoparticles with the skin barrier: current status and critical review. *Nanomed. Nanotechnol. Biol. Med.*, 9, 1, 39–54, 2013. ISSN 1549-9634, https://doi.org/10.1016/j.nano.2012.04.004.
22. Lademann, J., Weigmann, H.-J., Rickmeyer, C., Barthelmes, H., Schaefer, H., Mueller, G., Sterry, W., Penetration of titanium dioxide microparticles in a sunscreen formulation into the horny layer and the follicular orifice. *Skin Pharmacol. Appl. Skin Physiol.*, 12, 247–256, 1999.
23. Lademann, J., Richter, H., Teichmann, A., Otberg, N., Blume-Peytavi, U., Luengo, J., *et al.*, Nanoparticles—An efficient carrier for drug delivery into the hair follicles. *Eur. J. Pharm. Biopharm.*, 66, 159–64, 2007.
24. Lee, S.H., Lee, H.R., Kim, Y.R., Kim, M.K., Toxic Response of Zinc Oxide Nanoparticles in Human Epidermal Keratinocyte HaCaT Cells. *J. Toxicol. Environ. Health Sci.*, 4, 14–18, 2012.
25. Meyer, K., Rajanahalli, P., Ahamed, M., Rowe, J.J., Hong, Y., ZnO nanoparticles induce apoptosis in human dermal fibroblasts *via* p53 and p38 pathways. *Toxicol. Vitro*, 25, 1721–1726, 2011.
26. Moghimi, H.R., Barry, B.W., Williams, A.C., Stratum corneum and barrier performance: A model lamellar structural approach, in: *Drugs—Cosmetics—Mechanisms—Methodology*, R.L. Bronaugh and M. HI (Eds.), pp. 515–53, Marcel Dekker, New York, Basel, Hong Kong, 1999.
27. Monteiro-Riviere, N.A. and Riviere, J.E., Interaction of nanomaterials with skin: aspects of absorption and biodistribution. *Nanotoxicology*, 3, 3, 188–193, 2009.
28. Monteiro-Riviere, N., Nemanich, R., Inman, A., Wang, Y., Riviere, J., Multi-walled Carbon nanotube interactions with human epidermal keratinocytes. *Toxicol. Lett.*, 155, 377–384, 2005.

29. Monteiro-Riviere, N.A., Structure and Function of Skin, in: *Toxicology of the Skin—Target Organ Series*, N.A. Monteiro-Riviere, (Ed.), pp. 1–18, Informa Healthcare, Raleigh, NC, USA, 2010, ISBN 9781420079173.
30. Moss, G.P., Dearden, J.C., Patel, H., Cronin, M.T., *Toxicol in Vitro*, 16, 299–317, 2002.
31. Nabeshi, H., Yoshikawa, T., Matsuyama, K., Nakazato, Y., Arimori, A., Isobe, M., Tochigi, S., Kondoh, S., Hirai, T., Akase, T., *et al.*, Amorphous nanosilicas induce consumptive coagulopathy after systemic exposure. *Nanotechnology*, 23, 045101, 2012.
32. Oesch, F., Fabian, E., Oesch-Bartlomowicz, B., Werner, C., Landsiedel, R., *Drug Metab. Rev.*, 39, 659–698, 2007.
33. Park, Y.H., Bae, H.C., Jang, Y., Jeong, S.H., Lee, H.N., Ryu, W.I., Yoo, M.G., Kim, Y.R., Kim, M.K., Lee, J.K., *et al.*, Effect of the size and surface charge of silica nanoparticles on cutaneous toxicity. *Mol. Cell. Toxicol.*, 9, 67–74, 2013.
34. Pernodet, N., Fang, X., Sun, Y., Bakhtina, A., Ramakrishnan, A., Sokolov, J., Ulman, A., Rafailovich, M., Adverse effects of citrate/gold nanoparticles on human dermal fibroblasts. *Small*, 2, 766–773, 2006.
35. Rao, J.P. and Geckeler, K.E., Polymer nanoparticles: Preparation techniques and sizecontrol parameters. *Prog. Polym. Sci.*, 36, 7, 887–913, 2011.
36. Ryman-Rasmussen, J.P., Riviere, J.E., MonteiroRiviere, N.A., *Toxicol. Sci.*, 91, 1, 159–165, 2006.
37. Saweres-Argüelles, C., Ramírez-Novillo, I., Vergara-Barberán, M., Carrasco-Correa, E.J., Lerma- García, M.J., Simó-Alfonso, E.F., Skin absorption of inorganic nanoparticles and their toxicity: A review. *Eur. J. Pharm. Biopharm.*, 182, 128–140, 2023. doi: 10.1016/j.ejpb.2022.12.010.
38. Sayes, C., Liang, F., Hudson, J., Mendez, J., Guo, W., Beach, J., Moore, V., Doyle, C., West, J., Billups, W., Ausman, K., Colvin, V., Functionalization density dependence of single-walled carbon nanotubes cytotoxicity *in vitro*. *Toxicol. Lett.*, 161, 135–142, 2006.
39. Sayes, C., Wahi, R., Kurian, P., Liu, Y., West, J., Ausman, K., Warheit, D., Colvin, V., Correlating nanoscale titania structure with toxicity: a cytotoxicity and inflammatory response study with human dermal fibroblasts and human lung epithelial cells. *Toxicol. Sci.*, 92, 74–185, 2006.
40. SCENIHR SCoEaN-IHR, Risk Assessment of Products of Nanotechnologies. *Health Consum. Prot.*, 2009.
41. Schneider, M., Stracke, F., Hansen, S., Schaefer, U.F., Review: Nanoparticles and their interactions with the dermal barrier. *Dermato-Endocrinology*, 1, 4, 197–206, 2009.
42. Schulz, J., Hohenberg, H., Pflücker, F., Gärtner, E., Will, T., Pfeiffer, S., *et al.*, Distribution of sunscreens on skin. *Adv. Drug Deliv. Rev.*, 54, S157–S163, 2002.
43. Scientific Committee on Cosmetic and Non-Food Products (SCCNFP), Opinion concerning Titanium Dioxide (Colipa n S75). SCCNP: Brussels, 2000, 2000.

44. Shvedova, A., Castranova, V., Kisin, E., Schwegler-Berry, D., Murray, A.R., Gandelsman, V.Z., Maynard, A., Baron, P., Exposure to carbon nanotube material: assessment of nanotube cytotoxicity using human keratinocyte cells. *J. Toxicol. Environ. Heal.*, 66, 1909–1926, 2006.
45. Sonavane, G., Tomoda, K., Sano, A., Ohshima, H., Terada, H., Makino, K., *In vitro* permeation of gold nanoparticles through rat skin and rat intestine: Effect of particle size. *Colloids Surf., B*, 65, 1–10, 2008.
46. Souto, E.B., Almeida, A.J., Müller, R.H., Lipid nanoparticles (SLN®, NLC®) for cutaneous drug delivery: Structure, protection and skin effects. *J. Biomed. Nanotechnol.*, 3, 317–31, 2007.
47. Sur, S., Rathore, A., Dave, V., Reddy, K.R., Chouhan, R.S., Sadhu, V., Recent developments in functionalized polymer nanoparticles for efficient drug delivery system. *Nano-Struct. Nano-Objects*, 20, 100397, 2019. https://doi.org/10.1016/.
48. Sykes, E.A., Dai, Q., Tsoi, K.M., Hwang, D.M., Chan Warren, C.W., Nanoparticle exposure in animals can be visualized in the skin and analysed *via* skin biopsy. *Nat. Commun.*, 5, 3796, 2014.
49. Tak, Y.K., Pal, S., Naoghare, P.K., Rangasamy, S., Shape-Dependent Skin Penetration of Silver Nanoparticles: Does It Really Matter? *Sci. Rep.*, 5, 16908, 2015.
50. Tan, M.-H., Commens, C.A., Burnett, L., Snitch, P.J., A pilot study on the percutaneous absorption of microfine titanium dioxide from sunscreens. *Aust. J. Dermatol.*, 37, 185–187, 1996.
51. Tinkle, S.S., Antonini, J.M., Rich, B.A., Roberts, J.R., Salmen, R., DePree, K., Adkins, E.J., Skin as a route of exposure and sensitization in chronic beryllium disease. *Environ. Health Perspect.*, 111, 1202–8, 2003.
52. Toll, R., Jacobi, U., Richter, H., Lademann, J., Schaefer, H., Blume-Peytavi, U., Penetration profile of microspheres in follicular targeting of terminal hair follicles. *J. Invest. Dermatol.*, 123, 168–76, 2004.
53. Watkinson, A.C., Bunge, A.L., Hadgraft, J., Lane, M.E., Nanoparticles do not penetrate human skin-A theoretical perspective. *Pharm. Res.*, 30, 1943–1946, 2013.

6

Different Exposure Scenarios and Nanotoxicological Information

The Industrial Revolution as a whole was not designed. It took shape gradually as industrialists and engineers figured out how to make things. The result is that we put billions of pounds of toxic materials in the air, water and soil every year and generate gigantic amounts of waste. If our goal is to destroy the world—to produce global warming and toxicity and endocrine disruption—we're doing great.

William McDonough

Abstract

This chapter reviews nanoparticle exposure across the entire lifecycle—from manufacturing to use and disposal—affecting workers and consumers in industries such as medicine, agriculture, and defense. Nanomaterials behave differently than bulk materials, requiring tailored risk assessments. The chapter outlines biokinetic pathways and factors influencing nanotoxicity, including size, morphology, surface properties, and solubility. It recommends a tiered approach: *in vitro* tests, followed by *in vivo* studies including histopathology, immunotoxicity, and organ-specific toxicity. Emerging tools like (Q)SAR models and AI-based risk prediction are also discussed.

Keywords: Biokinetic-pathway, toxicity, cyto-toxicity, nanotoxicity, immunotoxicity, genotoxicity, metabolomic, proteomic, histopathology

6.1 Introduction

Nanotechnology involving engineered nanoparticles (ENPs) has grown into a multibillion-dollar industry worldwide. The importance of ENPs and composites is envisaged as immense due to their unique physicochemical characteristics that can be used in various fields of medicine, pharmaceutics, biotechnology, energy production, environmental sciences, crop protection,

Madhuri Sharon. Risk, Safety, and Ethical Concerns in Nanotechnology, (103–138) © 2025 Scrivener Publishing LLC

transportation, housing, and electronics [1–3]. Though the potential toxic effects of ENPs are being noted due to their exposure to human beings and the environment, we still need more understanding about the interactions of ENPs with biological systems and the possible side effects. In the previous chapters we have noted that ENPs are able to enter the body through various routes and can translocate to the bloodstream, cross body membrane barriers, blood–brain barrier (BBB), and eventually enter into various organs and tissues at cellular and molecular levels. Various levels of ENP interactions with proteins in the organs are different than those observed with conventional material, which has resulted in creating toxic effects. These effects vary due to size, distribution, dose, and duration of exposure based on the organs affected. In this chapter, we will be exploring exposure, biological pathways, toxic potential, and underlying mechanisms responsible for such toxic effects of ENPs. This will support the necessary formulations based on a toxicological risk assessment to human health during manufacture, use, and disposal.

One has to realize that the ENPs are manufactured using different chemical elements such as metals, metal oxides, sulfides or selenides, carbon, polymers, and biological molecules (lipids, carbohydrates, peptides, proteins, and nucleic acid oligomers). They all exhibit different geometry. For example, engineered nano forms of carbon itself are synthesized in different geometric configurations, i.e., isometric fullerene, one-dimensional carbon nanofibers, single or multi-wall carbon nanotubes, and two-dimensional plate or disk-like graphene sheets (Figure 6.1). Various shapes of gold NPs (Figure 6.2) have been synthesized, e.g., spherical [4], triangular [5], cube-like [6] and rod-like structures [7]. Since the diversity of engineered ENPs is not only in chemicals or size but also in shapes, their

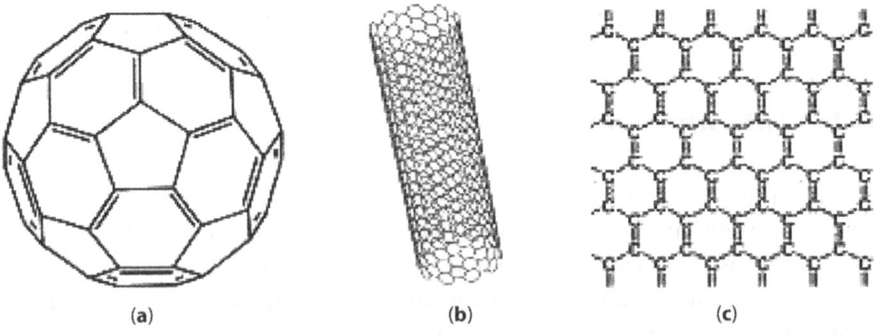

Figure 6.1 Nanoforms of carbon exhibiting different geometric configurations: (a) isometric fullerene, (b) one-dimensional single-wall carbon nanotube, and (c) two-dimensional plate or sheet of graphene.

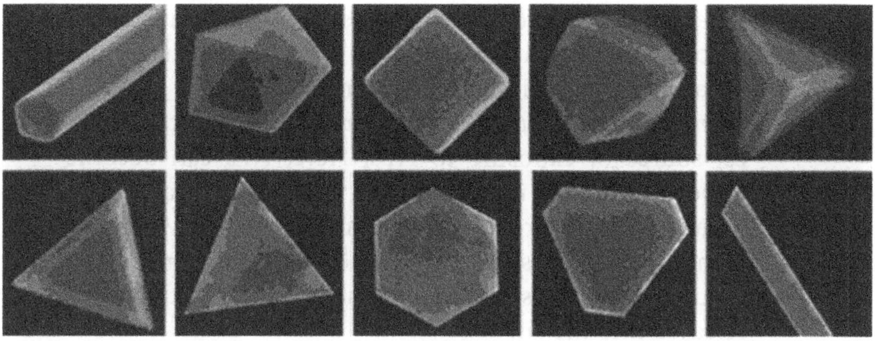

Figure 6.2 TEM images of various shapes of engineered gold NPs.

entry and interaction with living system are more complex, making it more difficult to formulate a standard exposure, toxicity, and risk model. The adverse effect of ENPs on human health can be during their manufacture, use or application, and finally related to possible excretion from the body. Nowadays there are precautionary measures taken during manufacturing. However, humans may be exposed to them in other ways due to their presence in the atmosphere.

Included in the biomedical applications for ENPs are therapeutics (drug delivery) and diagnostics.

Diagnostics: Many polymeric ENPs and quantum dots of silica and polymers are used for fluorescence imaging; the superparamagnetic iron oxide NPs used for MRI and the gold NPs used for photoacoustic imaging are biocompatible and their biosafety and biocompatibility have been reviewed [8]. Studies have shown that the human lung is one of the most easily and dangerously organs affected by ENP exposure [9]. The toxic effect of ENP has been assessed for many different ENPs based on different exposure scenarios and interactions of NPs with biological systems. However, the mechanisms of their toxic actions have yet to be assessed in detail, which is why a risk management process should be done.

6.2 Exposure Scenario to NPs

Exposure to NPs in various scenarios has become a great cause of concern as there are many steps starting with manufacturing and ending with post-disposal when the workers may get exposed. Therefore, scientists are involved in the following steps:

(i) *Research and Development:* In this core area of nanotechnology, scientists and researchers working with nanomaterials in laboratories face exposure risks.

(ii) *Manufacturing Industries:* These are bound to be the greatest source of exposure to NPs. Here, NPs can become airborne during manufacturing processes, leading to inhalation exposure for workers in the vicinity. Moreover, combustion processes, such as those seen in engines and industrial facilities, can also generate NPs that may be released into the air.

(iii) *Handling, Packaging and Transportation:* During synthesis or fabrication of ENP-related activities and nanoproducts, workers are sure to be exposed to NPs unless strict handling rules are followed. Accidental spills or releases during the transportation, handling, or processing of nanomaterials can result in unexpected exposure scenarios, hence demand a disaster management system for nanotechnology-related activities.

(iv) *Consumer Products:* Sunscreens, cosmetics, clothing, and food additives come in direct contact with users. Those using these products may be exposed to NPs through skin contact, inhalation, or ingestion.

(v) *Medical and Healthcare Products:* Products used in drug delivery systems, and as imaging agents, and diagnostic tools are bound to be a cause for concern for healthcare workers (doctors, nurses, hospital staff) and patients being treated with the use of NP-containing medical products.

(vi) *Disposal Process of NPs:* The disposal of nanoproducts, industrial effluents, and waste can release NPs into the environment. Workers involved in construction and demolition activities may be exposed to NPs present in construction materials or generated during these processes. Workers handling disposal of medicines, chemical or even blood and excreta of patients may face NPs in discarded products that can enter the environment through landfill sites, potentially impacting workers involved in waste management.

(vii) *Use of Nanomaterials in Agriculture:* Nanoparticles used in pesticides and fertilizers may lead to environmental exposure as well as direct exposure to farm workers.

(viii) *Nanoweapons:* These weapons, which are being developed under the cloak of secrecy by many countries, are expected to be more problematic than nuclear weapons. Because they are nearly invisible weapons of mass destruction, their destructive capacity and release of NPs will continue to affect the environment and living systems. Controlling millions of nanobots is difficult; once released, they may go beyond an adversary's boarders and kill indiscriminately [10].

It is important to note that the potential health effects of exposure to NPs is an active area of research, as these materials may behave differently than larger particles of the same chemical composition. Occupational health and safety measures, as well as environmental regulations, are evolving to address the challenges associated with nanomaterial exposure. Precautionary measures and risk assessments are crucial to minimizing potential risks associated with NP exposure.

6.3 Nanoparticle Exposure and Biokinetic Pathways

Biokinetics is a medically recognized discipline that uses scientifically based physical activity to prevent disease or rehabilitate patients after disease onset. Biokinetics is achieved using biokinetic model. A biokinetic model is a mathematical technique that predicts how materials are distributed in the body over time and how quickly they are eliminated through specific excretion pathways. It describes the time-dependent behavior of material that enters the body *via* a number of possible routes of intake and can also describe how different NPs are distributed and excreted after being absorbed into the blood or ingested by an organism. This model has been developed for uranium by the International Commission on Radiological Protection (ICRP), which predicts that 98% of natural uranium ingested through food or water is excreted in feces, while the remaining 2% enters the bloodstream.

We know that the NPs can enter the human body *via* inhalation, dermal, oral, and injection routes, either unintentionally or deliberately; however, to assess it using a biokinetic model, the important prerequisites that have to be considered for its successful use are the ability of NPs to cross biological barriers, such as the skin, gastrointestinal (GI) tract, or blood-brain barrier (BBB). Lungs are a different story, because inhalation is the most significant route of entry for exposures to NPs. Hayes and Bakand [11] have reviewed the toxicological perspectives of inhaled therapeutics and NPs.

6.3.1 Biokinetic Pathway of NPs *via* Inhalation

The route of inhaled NPs depends on their size; NPs of <1 μm and <100 nm penetrate deeply into the alveolar region of the lungs. The deeper the NPs penetrate, the more difficult it is for the body to remove them. There is not an adequate mechanism for their removal; hence, a longer amount of time will be required to eliminate them from the lung, causing greater adverse health effects, which may be due to particle-tissue and particle-cell interactions [12]. However, ultrafine particles of 1–5 μm, while traveling through the nose get deposited in the tracheobronchial region and those bigger than that, i.e., 5–30 μm, impact the nasopharyngeal region [13, 14]. Alveolar macrophage-mediated clearance system cannot remove nano-size (<100 nm) inhaled particles; and they tend to enter the bloodstream [15] and reach other target organs and even cross the BBB and enter the central nervous system (CNS). This is because they have the potential to cross the lung epithelial tissue barrier, unless they form aggregates or they are insoluble in water, in which case they may reside in the lung indefinitely. Figures 6.3 and 6.4 show the potential biokinetic pathway of inhaled NPs. Now the need is to (i) understand the clearance timeframes of NPs, (ii) standardize and develop the methods to characterize the NPs, (iii) quantify the NPs in biological systems by imaging techniques,

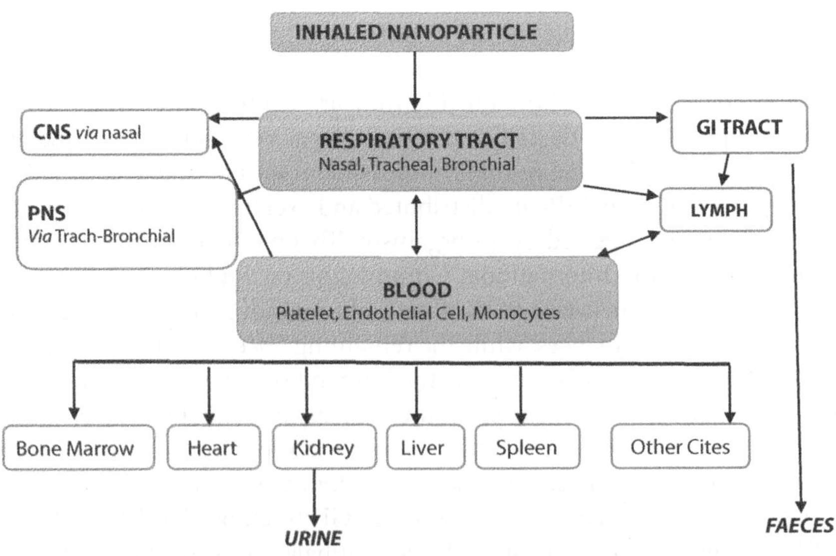

Figure 6.3 Biokinetics pathways of NPs (CNS: Central Nervous System; PNS: Peripheral Nervous System; GI: gastrointestinal tract).

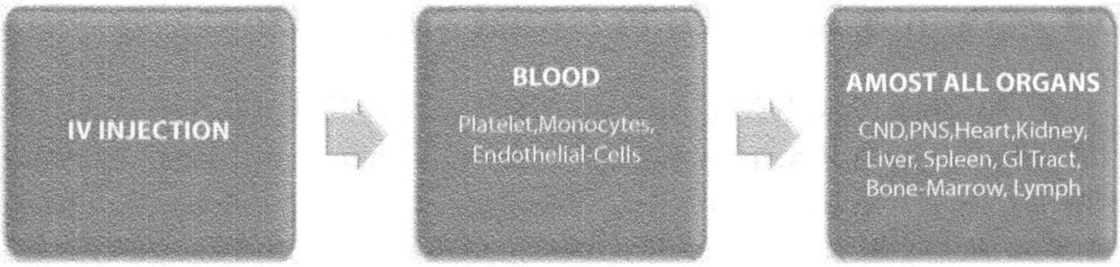

Figure 6.4 Biokinetics pathways of NPs (CNS: Central Nervous System; PNS: Peripheral Nervous System; GI: Gastrointestinal).

(iv) develop biomarkers of exposure, biomarkers of effects in living systems based on the interactions of NPs with target molecules, cells, and tissues, and (v) decide the appropriate sample by using toxicological data. The biological monitoring of inhaled nano-sized particles has been reviewed by Rinaldo *et al.* [9].

6.3.2 Biokinetic Pathway of Injected NPs

Injection of nanoscale therapeutics serves as an important route of deliberate entry of NPs. Injected NPs have easy and maximum access to all the organs through the blood circulatory system.

6.3.3 Biokinetics Pathway of Dermal Exposure to NPs

As compared to the inhaled or injected route of NPs, the dermal absorption or penetration studies have shown that NPs are neither able to easily penetrate intact skin nor through superficial injuries within the skin. A mechanical barrier is offered by the stratum corneum as well as secretion of ichor (a watery discharge from a wound or ulcer) and other blood components of the skin layer, which protects it from harmful impacts. For example, silver NP, is generally used due to its strong antibacterial, antiviral, antifungal and antimicrobial properties. It is found that subdermal administration of AgNPs induces oxidative stress and impairs the hepatic, renal and cardiac functions of male Wistar rats [16]. Such inconsistent data about dermal absorption and skin penetration of NPs suggests better evaluation is needed.

6.4 Nanotoxicity and Potential Mechanism

Many reports have surfaced over the last three decades associating potential risks of air polluting nanomaterials, especially from diesel engines, which release high levels of NPs (<50 nm) [17], causing increased mortality, respiratory illness, etc. [18]. Toxicity of NP is mainly associated with the following parameters.

6.4.1 Size and Size Distribution of NPs

As NPs cross through the tissue and cell membranes, they cause cellular injury and impaired phagocytosis, leading to migration to other organs and breakdown in defense mechanisms. Moreover, it may affect the transportation of other environmental pollutants. NPs <50 nm appear to enter cells and subcellular organelles, such as mitochondria and the nucleus, by passive diffusion [19] and cause direct cellular injury by particle-cell interactions [8]. If they enter the blood and CNS, they are bound to affect cardiac and cerebral functions [20, 21]. Aggregation of NPs affect their overall size, surface area, and surface charge, thereby influencing their cellular uptake. Aggregated NPs may have reduced cellular internalization due to increased particle size or altered surface properties. Preventing or controlling aggregation through proper surface modifications or dispersion techniques can help to improve NP uptake.

6.4.2 Morphology of NPs

The shape of NPs also plays a significant role in cellular uptake. Studies have shown that shapes, such as rods, disks, or spheres, may have different cellular uptake efficiencies. For example, elongated or rod-shaped NPs may have enhanced cellular internalization compared to spherical NPs due to their increased aspect ratio, which can facilitate interactions with cellular structures. Especially rod, triangular, tubular, fibers, or cube-like structure, which have high aspect ratio, impairs the phagocytic function of the alveolar macrophages, sometimes incomplete uptake, or frustrated phagocytosis [22]. Reduced capacity of removal of nanofibers by microphages from the lungs is reported [23]. Almost all nanoforms of carbon from carbon dots to SWCNT (single-walled carbon nanotube) has been found to have an adverse effect on human bio-system. Carbon dots easily translocate to other organs [21], triggering inflammatory or immunological responses [21].

Whereas SWCNTs translocate from the alveoli into the interstitium of the lung, promoting collagen deposition and interstitial fibrosis [24].

6.4.3 Surface Properties Such as Surface Area (SA) and Mass Ratio (MR)

Nanoparticles with higher surface area-to-volume ratios tend to have increased interactions with cells, leading to enhanced uptake. Higher surface area provides more sites for molecular interactions and can promote binding to cell surface receptors or facilitate cellular internalization processes. Greater reactivity, adsorption properties resulting in increased inflammation and toxicity are induced due to enhanced SA/MR [23, 25]. Inhalation of NPs by animals have been found to lead to more severe inflammatory responses compared to larger particles of similar composition and mass [26]. The surface charge of NPs can influence their interaction with cell membranes. Positively charged NPs may have increased uptake due to electrostatic interactions with negatively charged cell membranes. However, excessively positive charge can lead to cytotoxicity or nonspecific interactions with biomolecules. On the other hand, negatively charged NPs may experience reduced cellular uptake but may offer better stability and biocompatibility.

6.4.4 Chemical Composition and Surface Characteristics of NPs

Surface functionalization of NPs with specific ligands or targeting moieties can enhance their uptake by promoting specific interactions with receptors or biomolecules on the cell surface. Functional groups like antibodies, peptides, or aptamers can be conjugated to NPs to improve their targeting to specific cell types or tissues. Chemistry of NPs are the main cause of ROS generation leading to oxidative stress and inflammation. Cytokine production, glutathione depletion, mitochondrial exhaustion, cellular injury, and protein and DNA damage are also caused by chemical composition of NPs [27–30]. Oxidative stress which is responsible for toxicity of NMs, is induced either through direct generation of ROS at the surface of NPs or indirectly by target cells following the internalization of NPs [31].

6.4.5 Insolubility or Low Water Solubility of NPs

Insolubility induces bioaccumulation inside the human cells, tissues, and lungs, resulting in long-term adverse potential effects. They tend to bio-persist, influencing environmental and biological toxicity. Water solubility is associated with the toxicity of NPs [32]. Carbon nanomaterials (CNMs) are extremely insoluble in water and therefore can bioaccumulate inside living systems for a longer amount of time; hence, they are likely to cause pleural mesothelioma [33, 34]. Whereas water-soluble NPs provoke different types of adverse effects causing toxicity.

6.4.6 Agglomeration/Aggregation of NPs

The agglomerates may form in solid state or in liquid suspension [31]. Bioaccumulated NPs tend to agglomerate and interrupt the cellular processes and causes cellular injury. Using Zeta potential, information about the electrostatic interactions between particles in colloidal systems is obtained. Lower zeta potentials indicate the possibility of leading to flocculation or aggregation of particles due to weaker repulsive forces. This property is a complicating factor in toxicity testing of NPs.

6.5 Assessment of Toxicity of Nanoparticles

Assessing cellular toxicity *in vivo* using animal models, methods such as toxicokinetics, pulmonary inflammation, and potential models of fibrosis; will require three to four decades of extensive testing at rather exorbitant cost [35]. Therefore, to assess the cellular toxicity and the potential adverse effects of NPs, *in-vitro* assays are commonly employed. This includes assessing viability, membrane integrity, apoptosis, oxidative stress, and inflammatory responses.

6.5.1 *In-Vitro* Toxicity Assay Method

An *in-vitro* toxicity test is done using both human and animal-based cellular systems or cell lines. The *in-vitro* assay methods considered the standard widely used approach for evaluating cellular toxicity are:

(i) *MTT Assay (3-(4,5-dimethylthiazol-2-yl)-2,5-diphenyl tetrazolium bromide):* This assay measures cell viability based on the reduction of MTT by mitochondrial dehydrogenases in viable cells. The formazan product formed can be quantified spectrophotometrically.

(ii) *LDH (Lactate Dehydrogenase) Release Assay:* LDH is a cytoplasmic enzyme released upon cell membrane damage. An increase in LDH in the culture medium indicates cell membrane disruption and is indicative of cytotoxicity

(iii) *AlamarBlue Assay:* This assay measures cell viability by assessing the metabolic reduction of AlamarBlue dye. The reduced form of the dye is fluorescent and can be quantified using a fluorometer.

(iv) *Propidium Iodide (PI) Staining:* Propidium iodide is a fluorescent dye that is excluded from viable cells but penetrates cells with compromised membranes. It is often used in flow cytometry to assess cell membrane integrity.

(v) *Annexin V-FITC/PI Apoptosis Assay*: This assay distinguishes between apoptotic and necrotic cells using fluorescently labeled Annexin V and propidium iodide. Flow cytometry is commonly used for analysis.

(vi) *ROS (Reactive Oxygen Species) Assays:* Nanoparticles are known to induce oxidative stress leading to the generation of ROS. DCFDA (2',7'-dichlorofluorescein diacetate) and dihydroethidium (DHE) are commonly used probes for assessing intracellular ROS levels.

(vii) *Comet Assay (Single Cell Gel Electrophoresis):* This assay detects DNA damage at the single-cell level. Cells embedded in agarose are subjected to electrophoresis, and damaged DNA appears as a "comet tail" under microscopy.

(viii) *Nuclear Staining (e.g., Hoechst 33342):* Fluorescent nuclear dyes are used to visualize changes in nuclear morphology, such as chromatin condensation, which is indicative of apoptosis.

(ix) *Cytokine Release Assays:* Measurement of cytokine release (e.g., IL-6, TNF-α) can provide insights into NP-induced inflammation and immune responses.

(x) *TEM (Transmission Electron Microscopy):* This is not a traditional assay but is a powerful technique for visualizing internalization of NPs and their effects on cell ultrastructure.

The practice of combining multiple assays is used to get a more comprehensive understanding of the potential toxic effects of NPs on cells.

Depending on the organs and tissues that are affected, different *in-vitro* methods have been developed for respiratory or pulmonary toxicology human-based cellular systems such as airway cells, lung cells or tissues, and target-specific endpoints [36]. Such methods will provide potential benefits using simpler, faster, and less expensive toxicity screening tools [37]. However, there is a need to validate *in-vitro* assay systems for toxicity testing of an expanding range of NPs. Most of the trials using *in-vitro* models have been done using lung tissues exposed to airborne NPs.

6.5.2 *In-Vivo* Toxicity Assay Method

In-vivo assay methods for evaluating toxicity of NPs involve living organisms or animals such as mice, rats, rabbits or even humans. Animal studies involve exposing test animals to NPs and then observing their physiological responses such as changes in behavior, weight loss, organ damage, and mortality rates. These studies are supported by various analysis techniques of assaying status of tissues, blood, immune system, protein profile, DNA, reproductive system, various organs, etc.

6.5.2.1 Histopathological Analysis

Histopathological analysis involves examining tissue samples from animals exposed to NPs under a microscope. This allows researchers to identify any structural changes or abnormalities in organs or tissues, providing insights into the potential toxic effects of NPs. It involves analyzing and interpreting the shapes, sizes, and architectural patterns of cells and tissues; and it can even indicate the type of cancer; assess bone formation and resorption (rate of turnover or remodeling) and determination of the presence or absence of a superimposed mineralization defect. Histopathological testing plays a crucial role in assessing toxicity in various contexts, particularly in preclinical and clinical studies of NPs and environmental exposures. It is used in toxicity assessment of tissues under a microscope to identify any structural abnormalities or changes caused by toxicity. This includes looking for signs of inflammation, necrosis, apoptosis, fibrosis, hyperplasia, or hypertrophy in different organs and tissues. Histopathological analysis can provide insights into dose-response relationships, helping determine the severity of toxicity at different exposure levels to NPs. This information is essential for establishing dose guidelines and safety margins. Histopathology also allows for the assessment of temporal patterns of toxicity. By examining tissues at different time points after exposure, researchers can observe the progression of toxic effects and understand the dynamics of tissue response. Thus, it can provide mechanistic insights into the underlying processes of toxicity. For example, it can help elucidate whether toxicity is due to direct tissue damage, immune-mediated reactions, or other mechanisms. Histopathology enables comparative toxicology studies, where the effects of NPs of different types, size, and structure on tissues are compared. This information aids in understanding the relative toxicity of different NPs and guiding regulatory decisions. It can also help in safety assessment and environmental risk assessment, and helps identify potential hazards and informs decision-making regarding the safety of substances for human or environmental exposure. Moreover, histopathological findings can have predictive value for human health risks. Patterns of toxicity observed in animal studies can provide insights into potential adverse effects in humans and help guide risk management strategies.

6.5.2.2 Blood Chemistry Analysis

Blood chemistry analysis is a valuable tool in assessing NP toxicity, particularly in preclinical and clinical studies. Blood chemistry analysis involves measuring various biochemical parameters in the blood of animals exposed to NPs, such as liver enzymes, kidney function markers, and inflammatory markers. The toxicity of NP can affect hematological parameters, including red blood cell count, white blood cell count, and platelet count. Blood chemistry analysis can provide information on these parameters, aiding in the assessment of hepatotoxicity and potential effects on blood cell production and function. In the context of NP toxicity, the blood chemistry analysis is used for organ function assessment, which allows for the assessment of various biochemical markers related to organ function, e.g., levels of liver enzymes (e.g., ALT, AST) and kidney function markers (e.g., creatinine, blood urea nitrogen). These can indicate potential hepatotoxicity or nephrotoxicity induced by NP exposure. Exposure to NPs can elicit an inflammatory response in the body. Blood chemistry analysis can measure markers of inflammation, such as C-reactive protein (CRP) and pro-inflammatory cytokines, providing insights into the systemic inflammatory effects of NP toxicity.

As previously discussed, NPs may induce oxidative stress, leading to the generation of reactive oxygen species (ROS) and oxidative damage to cells and tissues. Blood chemistry analysis can assess markers of oxidative stress, such as lipid peroxidation products (e.g., malondialdehyde), antioxidant enzymes (e.g., superoxide dismutase, catalase), and antioxidant molecules (e.g., glutathione), helping to evaluate the oxidative stress response induced by NPs. Blood chemistry analysis allows for the evaluation of systemic effects of NP toxicity beyond specific organ toxicity. Changes in blood glucose levels, lipid profiles, electrolyte balance, and other metabolic parameters can indicate systemic disturbances induced by NP exposure. Since dose-response relationships are an important parameter in assessing the toxicity caused by NPs; blood chemistry analysis can help establish these relationships by measuring changes in biochemical markers at different exposure levels. This information is essential for determining the toxicity threshold and assessing the safety profile of NPs.

Blood chemistry analysis enables longitudinal monitoring of toxicity effects over time. By repeatedly assessing biochemical markers in blood samples, researchers can track the progression of toxicity, observe potential delayed effects, and evaluate the reversibility of adverse effects following NP exposure. Overall, blood chemistry analysis is a valuable complementary tool to histopathological assessment in evaluating NP toxicity. It provides quantitative data on systemic effects, organ function, inflammation, oxidative stress, and hematological parameters, enhancing our understanding of the toxicity mechanisms and safety profile of NPs.

6.5.2.3 Immunotoxicity Assessment

Immunotoxicity assessment plays a crucial role in understanding the potential health risks associated with NP exposure and informing decision-making regarding their safe use in various applications. It involves evaluating the impact of NPs on the immune system of animals and humans, such as assessing changes in immune cell populations, cytokine levels, and immune responses to pathogens or vaccines. Immunotoxicity assessment is crucial in analyzing NP toxicity for several reasons. Nanoparticles can interact with the immune system directly or indirectly, leading to immune responses that could be harmful or beneficial. Immunotoxicity assessment helps to understand these interactions and their potential consequences. Since NPs can induce inflammation, which may lead to tissue damage or exacerbate existing health conditions, immunotoxicity assessment helps evaluate the extent of these inflammatory responses and their impact on overall health. Some NPs can trigger allergic reactions or hypersensitivity responses in susceptible individuals. Immunotoxicity assessment helps identify potential allergenic properties of NPs and assess the risk of allergic reactions. Furthermore, NPs may also have immunosuppressive effects, compromising the body's ability to defend against pathogens or malignancies. Immunotoxicity assessment helps determine the extent of immunosuppression and its implications for health. Moreover, NPs have been implicated in triggering autoimmune responses, where the immune system mistakenly targets the body's own tissues. Immunotoxicity assessment helps evaluate the risk of autoimmune diseases associated with NP exposure. Understanding how NPs interact with immune cells is important as it can provide insights into their biodistribution and clearance from the body. Immunotoxicity assessment helps determine whether NPs accumulate in specific tissues or organs, potentially causing localized immune responses or toxicity. By integrating immunotoxicity data into risk assessments, regulatory agencies and researchers can better evaluate the safety of NP-based products and guide appropriate risk management strategies.

6.5.2.4 Genotoxicity Testing

Genotoxicity testing involves assessing the potential of NPs to damage DNA or induce mutations. This can be done using techniques such as comet assay, micronucleus assay, or DNA damage assays in animal cells or tissues. It a set of *in-vitro* and *in-vivo* tests that identify the direct or indirect effect of NPs causing genetic damage. Nanoparticles can interact with cellular DNA, leading to various forms of damage such as strand breaks, crosslinking, or base modifications. Genotoxicity testing helps assess the potential of NPs to induce DNA damage, which can lead to mutations and ultimately contribute to cancer development. This test can evaluate whether NPs have mutagenic effects, meaning they can cause changes in the genetic material of cells. Mutations induced by NPs may lead to altered gene expression, aberrant cell function, and potentially carcinogenesis. There are chances that many NPs can generate reactive oxygen species (ROS) as a byproduct of their interactions with cellular components. This ROS can cause oxidative stress and subsequent DNA damage. Genotoxicity testing helps assess the ability of NPs to generate ROS and the resulting impact on cellular DNA integrity. Moreover, this test can also evaluate whether exposure to NPs leads to cellular transformation that initiates cancer development. The cellular transformation induced by NPs may pose significant health risks and should be thoroughly evaluated for their genotoxic potential. Genotoxicity testing provides critical data for risk assessment and regulatory decision-making regarding the safe use of NPs in various applications, and thus helps to guide the development of appropriate risk management strategies. It allows for the comparison of different NP or NP-formulations to assess their relative safety profiles. It is valuable for regulatory agencies to prioritize the development and use of NPs with lower genotoxic potential.

6.5.2.5 Reproductive and Developmental Toxicity Studies

Development and reproductive toxicity (DART) studies, are used to assess the effects of drugs on a person's reproductive cycle. These studies assess the effects of NPs on reproductive health and development in animals. This may involve exposing pregnant animals to NPs and evaluating parameters such as fertility, pregnancy outcomes, and offspring development. The NPs may have adverse effects on reproductive organs and processes, potentially leading to impaired fertility in both males and females. Studies involve evaluating parameters such as sperm quality, ovulation, and reproductive hormone levels. The other problem could be that NPs can cross the placental barrier and reach the developing fetus, posing risks of embryotoxicity and teratogenicity. Developmental toxicity studies assess the effects of NP exposure on embryo development and fetal growth, including the incidence of structural abnormalities and developmental delays. The NPs may also interfere with the endocrine system, disrupting hormonal balance and signaling pathways critical for reproductive and developmental processes.

Reproductive and developmental toxicity studies help identify potential endocrine-disrupting effects of NPs and their implications for reproductive health. Moreover, exposure during critical periods of development may have long-lasting effects that may extend beyond the exposed generation. Reproductive toxicity studies can assess transgenerational effects by examining the reproductive and developmental outcomes in offspring of exposed animals across multiple generations. These analysis data on DART will contribute to assess the overall risk, and help in regulatory decision-making regarding their safe use for pregnant women, infants, and children. The data will also help in establishing safe exposure levels, guiding risk management strategies and regulatory guidelines. By determining thresholds for adverse effects on reproductive and developmental outcomes, these studies support the establishment of occupational exposure limits and environmental regulations. This information is critical for ensuring the safe development and use of NP-based products in various industries, including healthcare, cosmetics, and consumer goods.

6.5.2.6 Organ-Specific Toxicity Assessment

Nanoparticles may cause organ toxicity by acting to directly or indirectly enter into an organ *via* disruption of the normal physiological mechanisms affecting blood circulation. Therefore, some studies focus on evaluating the toxicity of NPs to specific organs such as the liver, kidneys, lungs, reproductive organs, or brain. Techniques such as organ weight measurements, imaging studies, and functional assessments can be used to evaluate organ-specific toxicity. For this, researchers identify specific organs or tissues that are likely to be exposed to NPs based on factors such as route of exposure, biodistribution, and known physiological interactions. The following assessments are done for this: (i) Histopathological examination; (ii) Functional assessment that measures parameters such as organ-specific biomarkers, enzyme activities, organ weights, and physiological endpoints relevant to the organ's function; (iii) Biodistribution studies to track the distribution and accumulation of NPs within the body to correlate exposure levels with organ-specific toxicity, and identify potential mechanisms of toxicity; (iv) Mechanistic studies to investigate the underlying mechanisms responsible for organ-specific NP toxicity. This may involve elucidating pathways of NP uptake, intracellular trafficking, cellular signaling, oxidative stress, inflammation, and other biological responses within target organs; and (v) Comparative analysis done to compares the toxicity of different NPs or NP-formulations to assess their relative impact on specific organs This information helps prioritize the development and use of NPs with lower organ-specific toxicity profiles.

These organ-specific NP toxicity assessments provide critical data for risk assessment and regulatory decision-making regarding the safe use of NPs in various applications. Organ-specific NP toxicity assessment is essential for comprehensively evaluating the safety profile of NPs and understanding their potential adverse effects on specific organs or tissues within the body.

6.5.2.7 Long-Term Exposure Studies

Long-term exposure studies involve exposing animals to NPs for extended periods to assess chronic toxicity effects such as carcinogenicity and cumulative toxicity over time. It involves exposing experimental animals or cell cultures to NPs for a significant portion of their lifespan, ranging from weeks to months or even years. The exposure period could be continuous or intermittent. Continuous exposure is done throughout the entire duration of the experiment whereas intermittent exposure regimens mimic real-world exposure scenarios. Intermittent exposure studies may assess the effects of repeated dosing schedules or intermittent exposure patterns relevant to occupational, environmental, or consumer settings. Long-term exposure studies involve careful selection of NP doses that reflect realistic exposure levels and allow for the detection of adverse effects over time. Doses may be chosen based on existing exposure data, occupational exposure limits, environmental concentrations, or dose-response relationships established in short-term toxicity studies. Throughout the duration of the study, health monitoring is done to assess the development of adverse effects associated with NP exposure. Monitoring may include clinical observations, body weight measurements, hematological and biochemical analyses, and histopathological examinations of tissues. Simultaneously, long-term exposure studies assess the effects of NP exposure on specific organs and physiological systems over time, such as histopathological examination, functional assays, and biomarker analyses to identify organ-specific responses to NP exposure. In addition to organ-specific effects, long-term exposure studies evaluate systemic effects of NP exposure, including effects on immune function, inflammatory responses, oxidative stress, genotoxicity, reproductive function, and neurotoxicity. Data collected from long-term exposure studies are analyzed to identify trends, dose-response relationships, and potential adverse effects associated with NP exposure. Results are interpreted in the context of existing toxicological knowledge, regulatory guidelines, and risk assessment frameworks to inform decision-making regarding NP safety. Long-term NP exposure studies are essential for comprehensively evaluating the potential health risks associated with prolonged exposure and for ensuring the safe development and use of NP-based products in various fields, including medicine, consumer products, and environmental applications.

6.5.2.8 Behavioral Studies

Behavioral studies are conducted to assess the impact of NP on cognition, neurological function, and related endpoints in experimental behavior in animals. This may involve tasks such as maze tests, motor function tests, and cognitive assessments. Experimental models used for behavioral studies include rodents (e.g., mice, rats), zebrafish, nematodes (e.g., *Caenorhabditis elegans*), and *in-vitro* neuronal cell cultures. Each model offers unique advantages for studying different aspects of behavior and neurological function in response to NP exposure. Behavioral assays assess a wide range of behaviors, including locomotor activity, anxiety-like behavior, depression-like behavior, cognitive function, learning and memory, social interaction, and sensorimotor coordination. Behavioral studies evaluate the impact of NP exposure on neurological function, such as synaptic transmission, neuronal excitability, neurotransmitter levels, neurogenesis, and synaptic plasticity. These assessments help identify potential mechanisms underlying NP-induced neurotoxicity. Cognitive function focuses on behavioral studies assessing the impact of NPs on learning, memory, attention, and executive function. Cognitive assays, such as the Morris water maze, novel object recognition, Y-maze, and radial arm maze, are commonly used to evaluate cognitive performance in NP-exposed animals. Behavioral studies may also investigate the effects of NP exposure during critical periods of neurodevelopment, including prenatal and early postnatal stages. These studies assess neurodevelopmental endpoints, such as neuronal migration, synaptogenesis, dendritic arborization, and myelination, and their implications for long-term neurological function. Behavioral studies also examine the role of neuroinflammation and oxidative stress in mediating NP-induced neurotoxicity. Neuroinflammatory markers, oxidative stress biomarkers, and cytokine levels in brain tissues are assessed to elucidate the underlying mechanisms of NP-induced behavioral changes. Moreover, long-term effects and dose-response relationships to determine the threshold levels of NP exposure that elicit behavioral changes, help to identify safe exposure levels, and inform risk assessment and regulatory decision-making. These studies contribute to our understanding of the potential health risks associated with NP exposure and support the development of strategies to mitigate adverse effects on neurological health.

6.5.2.9 Metabolomics and Proteomics

Metabolomics and proteomics are commonly used in NP toxicity studies to understand the biochemical and molecular changes induced by NP exposure. Metabolomics and proteomics techniques can be used to analyze changes in metabolite or protein profiles in animals exposed to NPs. These approaches can provide insights into the molecular mechanisms underlying NP toxicity.

Proteomics studies the proteins that make up an organism and their interactions. Proteomics also studies the catalysts of reactions, while metabolomics studies their substrates, intermediates, and products. Proteomics is the large-scale study of proteins, including their expression, structure, function, and interactions within biological systems. In NP toxicity studies, proteomics is used to identify changes in protein expression, post-translational modifications, and protein-protein interactions induced by NP exposure. Proteomics can provide valuable information about the molecular pathways and cellular processes affected by NPs, helping to elucidate their mechanisms of toxicity. Proteomic analysis can also identify potential protein biomarkers of NP exposure and toxicity, which can be used for risk assessment and the development of targeted therapeutic interventions.

Metabolomics studies the chemical fingerprints created from cellular processes. It is concerned with the substrates, intermediates, and products of reactions, and involves the comprehensive analysis of small molecule metabolites within biological systems. In NP toxicity studies, metabolomics is used to identify changes in metabolic pathways and to characterize the metabolic profiles associated with NP exposure. By analyzing changes in metabolite levels, metabolomics can provide insights into cellular responses, metabolic dysregulation, and the mechanisms underlying NP-induced toxicity. Metabolomics can also help identify biomarkers of NP exposure and toxicity, which can be used for early detection and risk assessment.

Both metabolomics and proteomics offer complementary information about the molecular and biochemical changes associated with NP exposure, providing a comprehensive understanding of NP-induced toxicity at the cellular and molecular levels. Integration of metabolomics and proteomics data with other omics technologies, such as genomics and transcriptomics, allows for a more holistic and systems-level approach to studying NP toxicity and its implications for human health.

It is important to note that ethical considerations should always be taken into account when conducting *in-vivo* studies, and efforts should be made to minimize animal suffering and reduce the number of animals used through appropriate experimental design and refinement of techniques. Additionally, regulatory guidelines and principles of good laboratory practice should be followed to ensure the reliability and reproducibility of study results.

With higher standardization, *in-vitro* methods can be precisely controlled to get more reproducible toxicity data than *in-vivo* models.

6.6 Risk Management of NPS

A risk is only relevant when there is a likelihood of exposure. Depending on the interaction with nanomaterials, the most likely routes of exposure may differ. Nanomaterials can enter the human body either orally *via* food intake, by penetrating the skin, or inhaled as part of the ambient air. The risk of nanomaterials is based on the identified toxicity (hazard characterization) of a chemical substance or nanomaterial, and the likelihood of exposure concentrations in humans and the environment. For risk characterization and assessment: (i) Hazard assessment is done using *in vitro, in vivo,* and epidemiological methods (epidemiology is the study and analysis of the distribution, patterns and determinants of health and disease conditions in a defined population); and (ii) Exposure assessment is done by measuring concentrations, by epidemiology studies, and computational modeling such as QSAR. Risk management of NPs is a big challenge, because of the vast diversity of ENMs as compared to other chemicals; be it size, shape, properties, chemical composition, or concentration. There are many ways in which NPs can interact with the surrounding compartments. Because of this, both the direct and indirect effects have to be considered. Another problem with NPs is that they can get dissolved in the water or sludge from wastewater treatment plants, soil, surface water, other pollutants, and constituents of the environment (e.g., oxides, silicates, phosphates, organic matter), and as a result become transformed or no longer exist as particulate material. Or get attached to other substances or degrade into another type of nano or micro particle, which may have more severe effects than the original nanomaterials, thus creating difficulties in quantifying or measuring the relevant concentrations of NPs to which organisms are exposed in the environment. These transformed nanomaterials.

However, the International Organization for Standardization (ISO) has published several standards and guidelines related to nanotechnology since 2008. These standards or guidelines cover various aspects, including terminology, measurement techniques, health and safety considerations, and environmental impacts. They also provide a framework to identify hazards, evaluate risk accordingly, and offer decision options, along with the risk communication procedures of manufactured NMs in order to protect the health and safety of the directly exposed population of workers, general public, and consumers through the environment. Following are the ISO series key ISO documents related to nanotechnology. These are just a few examples of ISO standards and technical reports related to nanotechnology.

1. **ISO/TS 80004 SERIES**, which provides terminology and definitions for nanotechnology. It includes:
   - ISO/TS 80004-1:2015 - "Nanotechnologies – Vocabulary – Part 1: Core terms"
   - ISO/TS 80004-2:2015 - "Nanotechnologies – Vocabulary – Part 2: Nano-objects"
   - ISO/TS 80004-3:2010 - "Nanotechnologies – Vocabulary – Part 3: Carbon nano-objects"
   - ISO/TS 80004-4:2011 - "Nanotechnologies – Vocabulary – Part 4: Nanostructured materials"
2. **ISO/TR 12885:2019** - "Nanotechnologies – Health and safety practices in occupational settings relevant to nanotechnologies."

- This technical report provides guidance on assessing and managing health and safety risks associated with nanomaterials in occupational settings.

3. **ISO/TS 12901 SERIES** - These documents provide guidance on the measurement of particle size distribution of nanomaterials in suspensions. It includes:
   - ISO/TS 12901-1:2012 - "Nanotechnologies – Particle size distribution and shape analysis of nanoscale particles in suspension by transmission electron microscopy"
   - ISO/TS 12901-2:2014 - "Nanotechnologies – Particle size distribution and shape analysis of nanoscale particles in suspension by scanning electron microscopy"
4. **ISO/TR 13121:2011:** "Nanotechnologies – Nanomaterial risk evaluation." This technical report provides guidance on the risk evaluation of nanomaterials.
5. **ISO/TR 16196:2014:** "Nanotechnologies – Guidance on methods for nano-ecotoxicity testing."
   This document provides guidance on ecotoxicity testing methods specific to nanomaterials.
6. **ISO/TR 16196:2014:** "Nanotechnologies – Guidance on methods for nano-ecotoxicity testing." This document provides guidance on ecotoxicity testing methods specific to nanomaterials.

Nanotechnology is a rapidly evolving field; hence a regular check of the ISO website is advisable along with other relevant sources for the most up-to-date information on standards and guidelines in this area.

6.6.1 Identification of Hazards Caused by Nanomaterials

Hazards may be due to both nanomaterial and its applications, which requires determining the profiles of NP, its physicochemical properties, and exposure details. Risk should be evaluated based on the combination of the identified hazards, exposure pattern, potential risks, and severity. Controlling risk at each level. At 1st Level, hazards due to NPs should be eliminated. At 2nd Level, consideration should be given to substituting a safer material, followed by controlling product or process as well as applying process containment during engineering of NP through use of local exhausted ventilation systems equipped with efficient HEPA filters. At 3rd Level, maximum exposure reduction should be imposed by the administrative controls so as to develop safety data sheets and safe work procedures.

Compulsory use of personal protective equipment, e.g., appropriate gloves, eye, and respiratory protection. Document and Act: it should be properly decided about whether or in what capacity to continue development and production of the NPs; all information should be collected and shared with the stakeholders. Finally, a serious Review is to be adapted regarding the update of the risk assessment process regularly and also the reviews triggered by specific events.

6.6.2 Guidance on Available Methods for Risk Assessment of Nanomaterials

In 2015, with support from the European Union's LIFE Fund in the framework of the LIFE Programme, the Environment Policy & Governance strand of LIFE+ entitled "LIFE REACHnano" developed a web-based REACH Toolkit to support the chemical safety assessment of nanomaterials" [38]. Its vision was to do a complete hazard assessment encompassing all the parameters that are discussed above in this subheading; Define the level of exposure and perform a thorough characterization to gain insight into levels of exposure and threshold levels below which risks for human health and for the environment occur. It has offered the best practice recommendations for risk assessment of nanomaterials.

Effects of naturally occurring NPs (NNPs) and man-made or engineered NPs (ENPs), taken up into an organism and environment was performed by Smita *et al.* [39], using an integrated approach to comprise available scientific information within an interdisciplinary logical framework, to identify knowledge gaps and to describe environment and health linkages for NNPs and ENPs. They have proposed a self-explanatory causal diagram as a method to handle the complexity of issues on NP safety, from their exposure to the effects on the environment and health. It gives an overview of available scientific information starting with common sources of NPs and their interactions with various environmental processes that may pose threats to both human health and the environment (Figure 6.5).

6.6.3 (Q)SAR Models for Regulatory Assessment of Nanomaterial Risks

Structure-activity relationship (SAR) and quantitative structure-activity relationship (QSAR) models—collectively referred to as (Q)SARs—are mathematical models. (Q)SAR models are computational models used in chemistry, drug discovery, environmental chemistry, toxicology, and

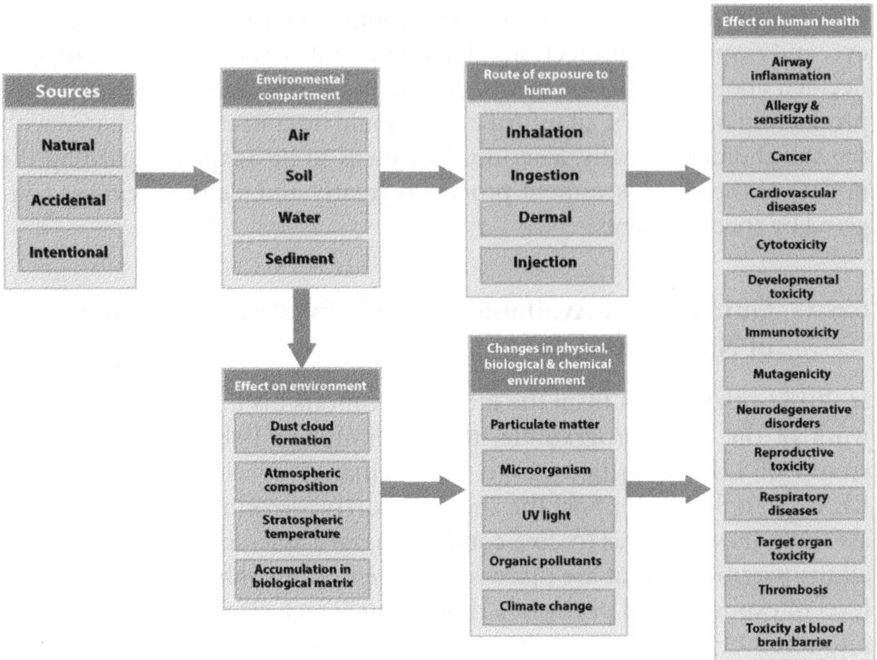

Figure 6.5 Proposed cause-effect diagram of NPs on environment and human health. (Source: Environmental Health [39]).

chemical safety assessment to predict the activity or properties of chemical compounds based on their structure. These models are based on the principle that the biological activity or physicochemical properties of a compound can be correlated with its molecular structure. They are valuable tools for prioritizing compounds for experimental testing, identifying potential hazards, and optimizing the design of new chemicals with desired properties. However, it is important to note that QSAR predictions are based solely on the structural information available and may not always accurately represent complex biological interactions or environmental conditions. Therefore, they are often used in conjunction with experimental data and other computational methods for comprehensive risk assessment and decision- making. The QSAR model development process involves data collection, calculation of molecular descriptors, activity/property measurement, model development, validation techniques, and finally, prediction. The question is can (Q)SAR models for regulatory assessment of nanomaterials risks be used? The answer is Yes, QSAR models can indeed be used for the regulatory assessment of nanomaterials risks, but there are several considerations and challenges to address when applying them in this context.

(i) *Data Collection:* QSAR models require a dataset of chemical compounds with known activities or properties obtained from experimental studies. These NP properties and activities are known and well studied when used. The QSAR models require large, high-quality datasets of nanomaterials with associated toxicity or physicochemical properties. However, such datasets may be limited for nanomaterials due to the novelty and complexity of these materials, as well as the challenges in standardizing characterization and testing methods.

(ii) *Molecular Descriptors:* These are quantitative representations of the molecular structure of compounds. The descriptors capture various aspects of a molecule's properties such as size, shape, electronic structure, and functional groups of NPs.

(iii) *Applicability Domain for NPs:* QSAR models are typically developed for specific chemical classes or endpoints. When applying QSAR models to nanomaterials, it is important to ensure that the models are applicable to the particular nanomaterials of interest and the relevant endpoints being assessed.

(iv) *Activity/Property Measurement:* The dataset is annotated with the measured biological activity or physicochemical properties of the compounds.

(v) *Model Development:* Using statistical and computational techniques, QSAR models are developed to establish relationships between the molecular descriptors and the activity or property of interest. Common methods include multiple linear regression, neural networks, support vector machines, and random forest.

(vi) *Model Validity and Reliability:* The performance of the QSAR model is assessed using validation techniques such as cross-validation or splitting the dataset into training and testing sets. This ensures that the model can generalize well to unseen data. The QSAR models need to be validated to ensure their accuracy, reliability, and robustness. For nanomaterials, validation may be more challenging due to the unique properties and behaviors of nanoscale materials, as well as potential interactions with biological systems that may not be fully captured by traditional QSAR approaches.

(vii) *Interpretation of Results:* Interpretation of QSAR model predictions for nanomaterials requires careful consideration of the underlying physicochemical properties and potential mechanisms of toxicity. Nanomaterials often exhibit size-dependent properties, surface reactivity, and complex interactions with biological systems, which may not be adequately captured by traditional QSAR descriptors.

(viii) *Prediction:* Once validated, the QSAR model can be used to predict the activity or property of new chemical compounds based solely on their molecular structure.

(ix) *Regulatory Acceptance:* Regulatory agencies may require evidence of the validity, reliability, and relevance of QSAR models for nanomaterials risk assessment before accepting them for regulatory purposes. This may involve collaborative efforts between researchers, industry stakeholders, and regulatory authorities to develop and validate QSAR models specifically tailored for nanomaterials. While QSAR models hold promise for the regulatory assessment of nanomaterial risks, their application requires careful consideration of data availability, model validity, interpretation of results, and regulatory acceptance. Integration of QSAR models with other experimental and computational approaches may provide a more comprehensive understanding of nanomaterials toxicity and support informed decision-making for regulatory purposes. Enrico Burello [40] has reviewed in detail the (Q)SAR models for regulatory assessment of nanomaterial risks.

6.6.4 Can AI Help in Risk Management of NPs

Artificial intelligence (AI) has the potential to enhance risk management practices related to NPs by enabling data-driven decision-making, predicting potential risks, ensuring regulatory compliance, and facilitating proactive mitigation strategies. However, it is essential to recognize the limitations of AI and ensure that human expertise is integrated into the risk management process to interpret results accurately and make informed decisions. AI is expected to play a significant role in risk management of NPs in the following ways:

6.6.4.1 *Analyzing Data and Predictive Modeling*

Artificial intelligence (AI) techniques, such as machine learning, can analyze large datasets to identify patterns and correlations between NP properties and their potential risks. By training models on historical data related to NP toxicity, environmental impact, and other factors, AI can predict the potential risks associated with new NPs using the following techniques: (i) Machine learning models using AI algorithms can analyze diverse datasets related to nanomaterials, including physicochemical properties, toxicity data, exposure levels, and biological responses. These algorithms can handle large volumes of data from various sources and integrate them to identify patterns, correlations, and trends. (ii) Nanomaterial datasets often contain a large number of features or descriptors, which can make modeling challenging. AI methods can automatically select relevant features or perform dimensionality reduction to reduce the complexity of the dataset while preserving essential information. (iii) AI techniques, including machine learning algorithms such as random forests, support vector machines, neural networks, and deep learning models, can be employed to develop predictive models for nanomaterial risk assessment. These models can predict various endpoints, such as toxicity, bioavailability, and environmental fate, based on the physicochemical properties of nanomaterials. (iv) AI models can provide insights into the relationships between nanomaterial properties and their potential risks. Techniques such as feature importance analysis and model visualization can help understand the underlying mechanisms driving nanomaterial toxicity and inform risk management strategies. (v) AI models can estimate the uncertainty associated with predictions, providing probabilistic assessments of nanomaterial risks. Uncertainty quantification is crucial for decision-making under uncertainty and can help prioritize resources and interventions effectively. (vi) AI models can be updated continuously as new data becomes available, allowing them to adapt to changing knowledge and emerging risks associated with nanomaterials. Continual learning techniques enable the incorporation of real-time data and feedback, improving the accuracy and reliability of risk predictions over time. (vii) AI models can be integrated into decision support systems for risk management of nanomaterials, providing stakeholders with actionable insights and recommendations. These systems can facilitate informed decision-making by synthesizing complex data and modeling results into accessible and interpretable formats.

6.6.4.2 *Toxicity Prediction*

AI models can be trained on toxicological data to predict the toxicity of NPs based on their physicochemical properties, such as size, shape, surface charge, and composition. These predictive models can help in assessing the potential health risks associated with exposure to specific NPs. AI can play a significant role in toxicity prediction for risk management of nanomaterials by leveraging various techniques and approaches: (i) *Machine Learning Models:* AI algorithms, models such as support vector machines (SVM), random forests, and neural networks can be trained on datasets containing information about the physicochemical properties of nanomaterials and their corresponding toxicity profiles. These models can learn complex patterns and relationships within the data to predict the toxicity of new nanomaterials. (ii) *QSAR Models:* Utilize AI techniques to correlate the chemical structure of nanomaterials with their toxicological properties. By analyzing the structure-property relationships, QSAR models can predict the toxicity of nanomaterials based on their molecular descriptors, such as size, shape, surface charge, and composition. (iii) *Processing and Analysis:* Vast amounts of data from diverse sources, including scientific literature, experimental studies, and databases containing toxicity information, can be processed and analyzed by AI. By integrating and analyzing this data using AI techniques, researchers can identify patterns and trends in nanomaterial toxicity, which can inform risk management strategies. (iv) *Deep Learning:* This is a subset of AI that uses artificial neural networks with multiple layers, has shown promise in predicting nanomaterial toxicity. Deep learning models can automatically extract relevant features from complex datasets and provide highly accurate toxicity predictions. Convolutional neural networks (CNNs) and recurrent neural networks (RNNs) are commonly used architectures in this context. (v) *Multi-Modal Data Fusion:* AI techniques can integrate data from multiple sources and modalities, including chemical properties, biological assays, and omics data (genomics, transcriptomics, proteomics, etc.), to improve toxicity predictions. By combining information from different data types, AI models can capture a more comprehensive understanding of nanomaterial toxicity and its underlying mechanisms. (vi) *Explainable AI (XAI):* Given the importance of interpretability in risk management, XAI techniques can provide insights into the decision-making process of AI models. By explaining the rationale behind toxicity predictions, XAI methods enhance the transparency and trustworthiness of AI-based toxicity prediction systems, enabling stakeholders to make informed decisions. (vii) *AI Models:* These can be designed to continuously learn and adapt to new data and emerging knowledge about nanomaterial toxicity. By incorporating feedback from experimental studies and real-world observations, AI systems can continuously improve their predictive accuracy and relevance for risk management. (viii) *Leveraging AI Techniques:* By leveraging these techniques researchers and regulatory agencies can enhance the prediction and management of nanomaterial toxicity, facilitating the safe and responsible development and use of nanotechnologies.

6.6.4.3 *Environmental Impact Assessment (EIA)*

Artificial intelligence (AI) algorithms can analyze data on the environmental fate and transport of NPs to assess their potential impact on ecosystems. By considering factors such as NP stability, mobility, and bioaccumulation potential, AI can help in identifying potential environmental risks and designing mitigation strategies. By leveraging AI technologies, environmental impact assessments for nanomaterials can be conducted more efficiently and accurately, enabling better risk management strategies to protect both the environment and human health. However, it is essential to ensure that these AI-driven approaches are transparent, interpretable, and based on high-quality data to foster trust and facilitate regulatory decision-making. Here's how AI can play a significant role in conducting EIAs for nanomaterials by providing advanced analytical capabilities and predictive modeling: (i) *Data Collection and Analysis:* AI algorithms can collect, process, and analyze vast amounts of data related to NPs, including their environmental fate and transport. This data can come from various sources such as scientific literature, databases, experimental studies, and simulations. (ii) *Predictive Modeling:* AI techniques such as machine learning and deep learning can be employed to develop predictive models that assess the environmental impact of NPs. These models can forecast the behavior of nanomaterials in different environmental compartments (e.g., air, water, soil), their bioaccumulation potential, and their ecological effects. (iii) *AI-Based Risk Assessment Frameworks:* These can be constructed to evaluate the potential risks associated with exposure to nanomaterials. These frameworks can consider factors such as exposure pathways, dose-response relationships, and environmental sensitivities to estimate the likelihood and severity of adverse effects on ecosystems and human health. (iv) *Decision Support Systems:* can be facilitated by AI that aid stakeholders in making informed decisions regarding the management and regulation of nanomaterials. These systems can provide recommendations based on the analysis of environmental impact assessments, risk assessments, and regulatory guidelines. (v) *Scenario Analysis:* AI algorithms can be used to assess the environmental impact of different usage scenarios of nanomaterials, such as manufacturing processes, product applications, and waste disposal methods. This helps in identifying potential hotspots of environmental risk and exploring mitigation strategies. (vi) *Continuous Monitoring and Adaptation:* AI-powered monitoring systems can continuously gather real-time data on the environmental presence and behavior of nanomaterials, enabling adaptive management strategies. These systems can detect emerging risks or unexpected environmental impacts and trigger timely responses to mitigate them. (vii) *Uncertainty Quantification:* AI techniques can be employed to quantify uncertainties associated with environmental impact assessments of nanomaterials. Probabilistic methods, Bayesian inference, and sensitivity analysis can be used to assess the robustness of assessment outcomes and provide insights into areas of uncertainty that require further research.

6.6.4.4 Regulatory Compliance

Artificial intelligence (AI) systems can assist in navigating the complex regulatory landscape surrounding NP risk management by automating compliance checks and providing real-time updates on regulatory changes. AI-powered tools can help companies ensure that their NP-based products meet regulatory requirements and standards. Moreover, AI technologies have the potential to revolutionize regulatory compliance assessment for nanomaterials by improving efficiency, accuracy, and agility in managing regulatory risks. However, it is essential to address challenges related to data quality, interpretability, and ethical considerations to ensure the responsible and effective use of AI in regulatory compliance management. AI can significantly aid in regulatory compliance assessment for risk management of nanomaterials by providing advanced tools for data analysis, decision support, and compliance monitoring. AI can be utilized in this context for (i) *Data Analysis and Documentation:* AI algorithms can streamline the process of collecting, organizing, and analyzing regulatory requirements related to nanomaterials from various sources such as government regulations, industry standards, and scientific literature. Natural language processing (NLP) techniques can be used to extract relevant information from regulatory texts and documents, enabling regulatory professionals to stay updated on evolving compliance requirements. (ii) *Compliance Monitoring and Auditing:* AI-powered monitoring systems can be used to track compliance with regulatory standards and requirements for the production, handling, and disposal of nanomaterials. These systems can analyze data from sensors, surveillance cameras, and other monitoring devices to detect deviations from regulatory guidelines in real time. Furthermore, AI algorithms can automate compliance audits by analyzing large volumes of data and identifying non-compliance patterns or anomalies. (iii) *Risk Assessment and Mitigation:* Addressed by AI techniques such as machine learning and predictive analytics to evaluate the potential hazards and risks associated with nanomaterials. By analyzing data on nanomaterial properties, exposure pathways, and toxicity profiles, AI models can predict the likelihood and severity of adverse effects on human health and the environment. This information can then be used to develop risk mitigation strategies and prioritize regulatory interventions. (iv) *Regulatory Intelligence and Strategy:* Should be provided by regulatory professionals with up-to-date information on changes in nanomaterial regulations and policies worldwide using AI-powered regulatory intelligence platforms. By analyzing regulatory trends, precedents, and enforcement actions, these platforms can help organizations anticipate regulatory developments and adapt their compliance strategies accordingly. Additionally, AI can assist in scenario analysis to evaluate the potential impact of regulatory changes on business operations and strategic decision-making. (v) *Compliance Training and Education:* Done through AI-driven e-learning platforms that can deliver personalized training and education programs on nanomaterial regulations and compliance requirements. These platforms can leverage adaptive learning algorithms to tailor training content to the specific needs and knowledge gaps of regulatory professionals. Furthermore, AI- based virtual assistants can provide on-demand guidance and support to users navigating complex regulatory frameworks and procedures. (vi) *Document Management and Reporting:* AI can automate the process of generating regulatory compliance reports and documentation for nanomaterials. Natural language generation (NLG) algorithms can convert structured data from compliance assessments and audits into comprehensive reports, summaries, and regulatory submissions. This streamlines regulatory reporting processes, reduces manual errors, and ensures consistency and accuracy in compliance documentation. (vii) *Decision Support Systems:* AI can be integrated into decision support systems for risk assessment and management, providing stakeholders with insights and recommendations for mitigating NP risks. These systems can consider various factors, including toxicity data, exposure levels, and risk tolerance thresholds, to inform decision-making processes. (viii) *Monitoring and Surveillance:* AI-enabled monitoring systems can continuously track the presence and behavior of NPs in various environments, such as workplaces, industrial settings, and natural ecosystems. By analyzing sensor data and other sources of information, AI can detect potential risks in real time and trigger appropriate responses such as containment measures or emergency protocols.

6.7 Conclusion

In conclusion, it can be said that it is necessary to investigate the current state of knowledge concerning the risks of NPs in the environment and human health and estimate whether current knowledge is sufficient to facilitate comprehensive and effective risk assessment of various NPs. For this, recommendations on future research in the field of risk assessment of NPs is to be provided. Because there are deep uncertainties that currently challenge and pervade every step of the risk assessment of NPs, the procedure is uncapable of properly serving its purpose. For each and every ENP, hazard assessment, dose-response assessment, exposure assessment, and risk characterization have to be taken into consideration.

References

1. Renn, O. and Roco, M., *White Paper on Nanotechnology Risk Governance*, International Risk Governance Council, Geneva, Switzerland, 2006.
2. Karn, B., Masciangioli, T., Zhang, W., Colvin, V., Alivisatos, P., *Nanotechnology and the Environment; Applications and Implications*, American Chemical Society, Washington, DC, USA, 2005.
3. Hougaard, K.S., Campagnolo, L., Chavatte-Palmer, P., Tarrade, A., Rousseau-Ralliard, D., Valentino, S., Park, M.V., de Jong, W.H., Wolterink, G., Piersma, A.H., *et al.*, A perspective on the developmental toxicity of inhaled NPs. *Reprod. Toxicol.*, 56, 118–140, 2015.
4. Kesarkar, R., Yeole, M., Dalvi, B., Madhuri, S., Chowdhary, A., Simplistic approach towards synthesis of highly stable and biocompatible L-cysteine capped gold nanosphere intermediate for drug conjugation. *Int. J. Pharm. Sci. Rev. Res.*, 31, 1Article No. 29, 143–146, 2015.
5. Dharmatti, R., Phadke, P., Mewada, A., Thakur, M., Pandey, S., Madhuri, S., Biogenic gold nano-triangles: Cargos for anticancer drug delivery. *Mater. Sci. Engineering C*, 44, 92–98, 2014. https://doi.org/10.1016/j.msec.2014.08.006.
6. Mewada, A., Pandey, S., Oza, G., Shah, E., Thakur, M., Gupta, A., Sharon Madhuri. A novel report on assessing pH dependent role of nitrate reductase on green biofabrication on gold nanoplates and nanocubes. *J. Bionanosci.*, 7, 1–7, 2013.
7. Pandey, G., Oza, G., Kesarkar, R., Gupta, A., Vishwanathan, M., Shah, R., Sharon, M., Madhuri Sharon, Taguchi Optimization method for tuning

aspect ratios for synthesis of gold nano rods;. *Arch. Appl. Sci. Res.*, 4, 2, 965–970, 2012.

8. Accomasso, L., Gallina, C., Turinetto, V., Giachino, C., Stem cell tracking with NPs for regenerative medicine purposes: An overview. *Stem Cells Int.*, 2016, 2016.
9. Rinaldo, M., Andujar, P., Lacourt, A., Martinon, L., Canal Raffin, M., Dumortier, P., Pairon, J.C., Brochard, P., Perspectives in biological monitoring of inhaled nanosized particles. *Ann. Occup. Hyg.*, 59, 669–680, 2015.
10. Madhuri, S., *Nanotechnology in Défense Industry*, John Wiley & Sons, Inc., 111 River Street, Hoboken, NJ 07030, USA, 2019.
11. Hayes, A.J. and Bakand, S., Toxicological perspectives of inhaled therapeutics and NPs. *Expert Opin. Drug Metab. Toxicol.*, 10, 933–947, 2014.
12. Blank, F., Gehr, P., Rutishauser, R.R., *In Vitro Human Lung Cell Culture Models to Study the Toxic Potential of NPs;*, Chichester, UK, John Wiley & Sons Ltd, 2009.
13. Siegmann, K., Scherrer, L., Siegmann, H.C., Physical and chemical properties of airborne nanoscale particles and how to measure the impact on human health. *J. Mol. Struct.*, 458, 191–201, 1999.
14. Witschi, H.P., Toxic responses of the respiratory system, in: *Casarett and Doull's Toxicology: The Basic Science of Poisons*, 6th ed, C.D. Klaassen, (Ed.), pp. 515–534, McGraw-Hill, New York, NY, USA, 2001.
15. Mublfeld, C., Gebr, P., Rutishauser, B.R., Translocation and cellular entering mechanisms of NPs in the respiratory tract. *Swiss Med. Wkly.*, 138, 387–391, 2008.
16. Olugbodi, J.O., Lawal, B., Bako, G., Onikanni, A.S., Abolenin, S.M., Mohammud, S.S., Ataya, F.S., Batiha, G.E., Effect of sub-dermal exposure of silver NPs on hepatic, renal and cardiac functions accompanying oxidative damage in male Wistar rats. *Sci. Rep.*, 13, 1, 11274, 2023, doi: 10.1038/s41598-023-37178-x.
17. Lawson, G. and Wang, H., Public health impact of diesel exhast: Toxicity of nano-sized diesel exhaust particles—Part 1. *Environ. Health*, 6, 17–21, 2006.
18. Borm, P.J., Robbins, D., Haubold, S., Kuhlbusch, T., Fissan, H., Donaldson, K., Schins, R., Stone, V., Kreyling, W., Lademann, J., *et al.*, The potential risks of nanomaterials: A review carried out for ECETOC.Part. *Fibre Toxicol.*, 2006, 3, 11.
19. Chithrani, D.B., Intracellular uptake, transport, and processing of gold nanostructures. *Mol. Membr. Biol.*, 27, 299–311, 2010.
20. Donaldson, K., Stone, V., Tran, C.L., Kreyling, W., Borm, P.J., Nanotoxicology. *Occup. Environ. Med.*, 61, 727–728, 2004.
21. Donaldson, K., Tran, L., Jimenez, L.A., Duffin, R., Newby, D.E., Mills, N., MacNee, W., Stone, V., Combustion-derived NPs: A review of their toxicology following inhalation exposure. *Part. Fibre Toxicol.*, 2005, 2, 10.

22. Shi, X., von dem Bussche, A., Hurt, R.H., Kane, A.B., Gao, H., Cell entry of one-dimensional nanomaterials occurs by tip recognition and rotation. *Nat. Nanotechnol.*, 6, 714–719, 2011.
23. Brown, D.M., Kinloch, J.A., Bangert, U., Windle, A.H., Walter, D.M., Walker, G.S., Scotchford, C.A., Donaldson, K., Stone, V., An *in vitro* study of the potential of carbon nanotubes and nanofibres to induce inflammatory mediators and frustrated phagocytosis. *Carbon*, 45, 1743–1756, 2007.
24. Wang, X., Xia, T., Ntim, S.A., Ji, Z., Lin, S., Meng, H., Chung, C.H., George, S., Zhang, H., Wang, M., *et al.*, Dispersal state of multiwalled carbon nanotubes elicits profibrogenic cellular responses that correlate with fibrogenesis biomarkers and fibrosis in the murine lung. *ACS Nano*, 5, 9772–9787, 2011.
25. Renwick, L.C., Brown, D., Clouter, A., Donaldson, K., Increased inflammation and altered macrophage chemotactic responses caused by two ultrafine particle types. *Occup. Environ. Med.*, 61, 442–447, 2004.
26. Warheit, D.B., NPs health impacts. *Mater. Today*, 7, 32–35, 2004.
27. Kim, I.Y., Joachim, E., Choi, H., Kim, K., Toxicity of silica NPs depends on size, dose, and cell type. *Nanomedicine*, 11, 1407–1416, 2015.
28. Khalili Fard, J., Jafari, S., Eghbal, M.A.A., review of molecular mechanisms involved in toxicity of NPs. *Adv. Pharm. Bull.*, 5, 447–454, 2015.
29. Nel, A., Xia, T., Madler, L., Li, N., Toxic potential of materials at the nanolevel. *Science*, 311, 622–627, 2006.
30. Karlson, H.L., Cronholm, P., Gustafsson, J., Moller, L., Copper oxide NPs are highly toxic: A comparison between metal oxide NPs and carbon nanotubes. *Chem. Res. Toxicol.*, 21, 1726–1732, 2008.
31. Oberdorster, G., Kane, A.B., Klaper, R.D., Hurt, R.H., Nanotoxicology, in: *Casarett and Doull's Toxicology—The Basic Science of Poisons*, C.D. Klaassen, (Ed.), pp. 1189–1229, McGraw Hill, New York, NY, USA, 2013.
32. Brunner, T.J., Wick, P., Manser, P., Spohn, P., Grass, R.N., Limbach, L.K., Bruinink, A., Stark, W.J., *In vitro* cytotoxicity of oxides NPs: Comparison to asbestos, silical, and the effects of particle solubility. *Environ. Sci. Technol.*, 40, 4374–4381, 2006.
33. Song, Y., Li, X., Du, X., Exposure to NPs is related to pleural effusion, pulmonary fibrosis and granuloma. *Eur. Respir. J.*, 34, 559–567, 2009.
34. Poland, C.A., Duffin, R., Kinloch, I., Maynard, A., Wallace, W.A., Seaton, A., Stone, V., Brown, S., Macnee, W., Donaldson, K., Carbon nanotubes introduced into the abdominal cavity of mice show asbestos-like pathogenicity in a pilot study. *Nat. Nanotechnol.*, 3, 423–428, 2008.
35. Choi, J.Y., Ramachandran, G., Kandlikar, M., The impact of toxicity testing costs on nanomaterial regulation. *Environ. Sci. Technol.*, 43, 3030–3034, 2009.
36. Bakand, S., Winder, C., Khalil, C., Hayes, A., Toxicity assessment of industrial chemicals and airborne contaminants: Transition from *in vivo* to *in vitro* test methods: A review. *Inhal. Toxicol.*, 17, 775–787, 2005.

37. Sayes, C.M., Reed, K.L., Warheit, D.B., Assessing toxicity of fine and NPs: Comparing *in vitro* measurements to *in vivo* pulmonary toxicity profiles. *Toxicol. Sci.*, 97, 163–180, 2007.
38. LIFE REACHnano -Development of a web based REACH Toolkit to support the chemical safety assessment of nanomaterials (LIFE11 ENV/ES/000549); REACHnano Consortium First Edition, Published in Spain, 2015.
39. Smita, S., Gupta, S.K., Bartonova, A., Dusinska, M., Gutleb, A.C., Qamar, R., NPs in the environment: assessment using the causal diagram approach. *Environ. Health*, 11, Suppl. 1, S13, 2012. http://www.ehjournal.net/content/11/S1/S1.
40. Burello, E., Review of (Q)SAR models for regulatory assessment of nanomaterials risks. *NanoImpact*, 8, 48–58, 2017.

7

Assessment of Risks Due to Nanoparticles in the Ecosystem

Waste does not exist in nature because ecosystems reuse everything that grows in a never-ending cycle of efficiency and purpose.

Frans van Houten

Abstract
This chapter shifts focus to the environmental risks of nanoparticles, which can impact air, water, and soil. While some nanoparticles occur naturally, most airborne particles—like carbon black, metal oxides, and organic compounds—are anthropogenic, originating from vehicles, industry, and agriculture. The chapter outlines how nanoparticles behave in ecosystems, stressing the need to assess their bioavailability, transport, and toxicity. It discusses current monitoring strategies, including remote sensing and tracking technologies, and calls for stakeholder collaboration—across governments, industry, science, and the public—to ensure effective ecosystem risk management.

***Keywords*:** Risk, nanoparticle, ecosystem, airborne NP, waterborne NP, bioavalability, transport of NP, fate of NP

7.1 Introduction

An ecosystem encompasses a biological community of interacting organisms and their physical environment. This includes all living organisms (biotic factors) in an area, such as plants, animals, fungi, and microorganisms, as well as non-living components (abiotic factors) such as soil, water, mineral, sunlight, and air. Ecosystems can vary greatly in size and complexity, ranging from a small pond to a vast rainforest. They are characterized by the flow of energy and the cycling of nutrients among the various components, which sustain life within the system. Ecosystems are dynamic

Madhuri Sharon. Risk, Safety, and Ethical Concerns in Nanotechnology, (139–162) © 2025 Scrivener Publishing LLC

and subject to change over time due to factors like climate, geological processes, human activities, and interactions among species. Examples of ecosystems include forests, grasslands, deserts, oceans, and coral reefs.

Ecosystems are not isolated entities; they are interconnected and interact with each other. For example, a forest ecosystem might provide habitat for a variety of organisms, from trees and birds to insects and fungi. These organisms interact with each other in complex ways, forming a delicate balance. Any change in one part of the ecosystem can have ripple effects throughout the entire system, highlighting the importance of conservation and understanding ecosystems for maintaining biodiversity and ecosystem services.

Ecosystems can be as large as a forest or as small as a pond. They can exist in various forms, including terrestrial ecosystems like deserts, grasslands, and forests, as well as aquatic ecosystems such as lakes, rivers, and oceans. Key components of an ecosystem include:

(i) *Producers:* These are typically plants or algae that produce their food through photosynthesis, converting sunlight into energy.
(ii) *Consumers:* Animals that eat other organisms. Consumers can be categorized into primary consumers (herbivores), secondary consumers (carnivores that eat herbivores), and tertiary consumers (carnivores that eat other carnivores).
(iii) *Decomposers:* Organisms like bacteria and fungi that break down dead organic matter, dead animal bodies and return nutrients to the soil or water. They play a crucial role in recycling nutrients in the ecosystem.
(iv) *Abiotic Factors:* These are the non-living components of an ecosystem, such as sunlight, temperature, water, soil, and nutrients. These factors greatly influence the types of organisms that can survive and thrive in a particular ecosystem.

Assessing the risk NP (NPs) pose to ecosystems requires a multidisciplinary approach and consideration of various factors. The guidelines on how to assess the risk posed by NPs to an ecosystem should, for example, evaluate effects related to novel physicochemical properties and stability of NPs. If exposures exceed innocuous levels in the ecosystem, there is a need to institute control measures and/or reduce the concentrations. Risk to ecosystem encompasses the presence of NPs in air, soil, and water [1].

(i) *Air:* Nanoparticles, often released by industrial emissions or combustion processes, enter the atmosphere. They can settle on plants, interfere with their growth, and affect the respiration and health of animals, such as birds. Over time, nanoparticles in the air can lead to the degradation of plant health and affect the food chain.

(ii) *Soil:* Nanoparticles infiltrate the soil through atmospheric deposition or wastewater. In the soil, they can disturb microorganisms critical for nutrient cycling, impact plant roots, and reduce soil fertility. These changes can ripple through the ecosystem, affecting insects, plants, and animals that depend on healthy soil.

(iii) *Water:* Nanoparticles can enter water bodies through runoff, industrial discharge, or even from air deposition. Once in water, they can harm aquatic organisms like fish, plants, and microorganisms. Nanoparticles can accumulate in the tissues of fish, disrupt aquatic food chains, and degrade water quality.

Each section of the ecosystem is connected by the flow and impact of nanoparticles, from their release to their effects on various organisms across ecosystems.

7.1.1 Airborne Nanoparticles

Airborne NPs are tiny particles suspended in the air that have dimensions typically <100 nanometers in diameter. They can be composed of various materials such as metals (like iron, nickel, or titanium), carbon (such as soot or carbon black), or organic compounds. These NPs can be of *natural origin*, such as those (i) produced by volcanic eruptions, (ii) produced by sea spray when waves break on the ocean's surface, and (iii) those released into the air by forest fires; or *anthropogenic sources* of NP generated by certain human activities, especially combustion processes, such as

(i) *Vehicle Emissions:* The combustion engines, especially diesel engines, produce NPs as a byproduct of incomplete combustion. The NPs from combustion can affect climate by influencing cloud formation and absorption or reflection of sunlight.

(ii) *Industrial Processes:* Factories and power plants release NPs during manufacturing processes and burning of fossil fuels. Nanoparticles, often released by industrial emissions or combustion processes, enter the atmosphere. They can settle on plants, interfere with their growth, and affect the respiration and health of animals, such as birds. Over time, nanoparticles in the air can lead to the degradation of plant health and affect the food chain.

(iii) *Construction Activities:* Cutting, drilling, and sanding materials like concrete or wood can create airborne NPs.

(iv) *Agricultural Activities:* Pesticides and fertilizers can produce NPs when sprayed or even added to the soil. The NPs can be deposited onto soil and water, affecting ecosystems, and potentially entering the food chain.

Nanoparticles are much smaller than the particles traditionally measured in air quality monitoring ($PM_{2.5}$ and PM_{10}), which are defined by their aerodynamic diameters of 2.5 and 10 micrometers, respectively. Due to their small size, airborne NPs can remain suspended in the atmosphere for extended periods and can be transported over long distances. They can also penetrate deep into the respiratory system when inhaled, potentially causing adverse health effects. The health impacts of airborne NPs depend on factors such as their composition, size, shape, surface properties, and concentration, as well as individual susceptibility and exposure duration. Since NPs have a large surface area compared to their volume, they are highly reactive [2]. Research on airborne NPs is ongoing to better understand their sources, behavior, distribution, and potential health and environmental effects. Strategies to mitigate exposure to airborne NPs include improving air quality through emission controls, developing personal protective equipment, and implementing regulations and policies to limit their release into the environment. The presence of NPs in air has been noted to occur individually or as aggregates. There are control options available for aggregates of NPs that are mostly of micrometer size. Whereas many NPs are of the size of gases or vapor molecules, hence risk assessment or exposure control may be employed as for gaseous matter.

It is challenging and difficult to provide an exact percentage of the world's area affected by heavy air pollution due to NPs as it varies significantly by region, time of year, and various other factors. Moreover, due to their small size and low mass, NPs can remain suspended in the air for extended periods, leading to long-distance transport. Air pollution levels can fluctuate due to industrial activities, traffic congestion, wildfires, agricultural practices, and natural geographical features, among other things. Due to their small size and potential health and environmental impacts, they are an important focus of air quality research and monitoring efforts. However, it is clear that air pollution is a widespread issue that requires global attention and concerted efforts to address.

7.1.2 Waterborne Nanoparticles

Waterborne NPs are dispersed or suspended in water. They can vary in size, shape, composition, and surface properties. Like airborne NPs, they can also be naturally occurring or engineered for various purposes. Naturally occurring NPs occur in water bodies due to geological processes, weathering of rocks and minerals, or biological activities. For example, colloidal particles, such as clays and silts, can exist in water as NPs. Additionally, organic matter in water bodies can form NPs through processes like microbial degradation. Whereas engineered NPs (ENP) are intentionally produced for specific applications such as in consumer products, industrial processes, and medical treatments. The ENPs can enter water bodies through various pathways, including industrial discharge, wastewater effluent, and runoff from NP-containing products that can be transported to aquatic systems by wind or rainwater runoff. They are a big concern due to their potential environmental and health impacts, as they can interact with aquatic organisms, potentially affecting their behavior, growth, reproduction, and overall ecosystem health. They may also accumulate in sediments, altering sediment characteristics and affecting benthic organisms. As far as the effect on human health is concerned, they can have adverse health effects, particularly through drinking water or recreational activities. Nanoparticles may penetrate biological barriers and tissues, leading to inflammation, oxidative stress, genotoxicity, and other health issues. For risk management, it is important to understand the fate, transport, transformation, and potential risks in aquatic environments. Strategies for mitigating their impact include improving wastewater treatment technologies, developing NPs with reduced environmental toxicity, and implementing regulations to limit their release into water bodies [3].

These concerns become more complex because NPs are being used to treat heavy metal contaminated water. For example, metallic oxide semiconductors like zinc oxide and titanium oxide can be used to remove heavy metal contamination from water. Some NPs, such as silver nanoparticles, are also used to disinfect water, as they are highly toxic to microorganisms and have strong antibacterial effects against a wide range of microorganisms, including viruses, and fungi. However, NPs in wastewater can have toxic effects on fish and crustaceans. Exposure to wastewater containing nanoparticles can lead to an increase in reactive oxygen species, which can react with and damage cells. A comparatively safer material to treat/filter wastewater is nanocellulose and carbon or cellulose nanofibrils (CNF), which are also used in water purification systems. In spite of diverse applications, they also pose environmental and health risks. Understanding their behavior, fate, and potential impacts is essential for managing water quality and protecting ecosystems and human health. Detecting and monitoring waterborne nanoparticles is rather challenging due to their small size and diverse composition. They need techniques, such as dynamic light scattering (DLS), electron microscopy (TEM and SEM), and atomic force microscopy (AFM), to visualize and characterize NPs in water. Advanced analytical methods like inductively coupled plasma mass spectrometry (ICP-MS) and single-particle inductively coupled plasma mass spectrometry (spICP-MS) are used for quantitative analysis.

7.1.3 Nanoparticles in Soil

Nanoparticles in soil can originate from both natural processes and human activities. Naturally occurring NPs in soil can arise from geological processes, weathering of rocks and minerals, and biological activities. Naturally occurring clay minerals in soils often consist of NPs. Organic matter decomposition and microbial activities also generate NPs in soil. Anthropogenic sources are human activities that can introduce NPs into soil through various pathways, such as agricultural practices, industrial activities, waste disposal, and the application of NP-based products (e.g., nanopesticides, nanofertilizers). For instance, engineered NPs used in consumer products may enter the soil through sewage sludge or runoff from treated surfaces. Nanoparticles in soil can have both beneficial and detrimental effects. Some NPs, particularly those derived from natural sources, can contribute to soil fertility, water retention, and nutrient availability. Engineered nanoparticles (ENPs) may also have applications in soil remediation, such as for contaminant immobilization or degradation. On the other hand, certain NPs may pose risks to soil organisms, plant health, and ecosystem functioning by interfering with soil microbial communities, affecting nutrient cycling processes, or inducing toxicity in soil-dwelling organisms. There are also concerns about the potential for NPs to migrate through soil and enter groundwater, posing risks to aquatic ecosystems and human health. Hence, thorough research to understand their behavior, fate, transformations, and potential ecological and human health impacts is needed to plan strategies for mitigating these risks. This may include developing sustainable nanomaterials, improving waste management practices, enhancing soil monitoring and remediation technologies, and implementing regulations to minimize NP exposure in soil environments. Environmental and agricultural impacts of NPs have been found to affect soil health, such as pH and nutrient availability, along with soil-dwelling organisms such as earthworms, microbes, and beneficial insects, and microbial activity. This can impact plant growth and soil fertility. Some NPs can be taken up by plants through their roots, potentially affecting plant growth and food quality. Studies have been conducted to assess the environmental risks associated with nanoparticles in soil, including their potential toxicity to organisms and their long-term effects on ecosystems [4]. In summary, nanoparticles in soil can have diverse origins and properties. While they have many potential applications, there are also concerns about their impact on soil health, plant growth, and environmental quality. Understanding the behavior and effects of nanoparticles in soil is essential for sustainable agriculture, environmental protection, and human health.

Case-by-case risk assessment of NPs present in all three components of environmental space and risk assessment of specific materials and their use patterns is needed.

7.2 What to Consider for Nanoparticle-Related Assessment of Risks to Ecosystem

Assessing the risks of nanoparticles to ecosystems is crucial for understanding their potential impacts on the environment. When assessing their risks, several factors should be considered to ensure a comprehensive evaluation. However, the first consideration should be to determine the specific NPs of concern that are being introduced into the ecosystem. This could include NPs used in consumer products, industrial processes, or those released unintentionally.

7.2.1 Identify and Characterize Properties of Nanoparticles of Concern

Nanoparticles have unique physicochemical properties due to their small size and high surface-area-to-volume ratio, which can influence their behavior and interactions within ecosystems. It is important to understand the physical and chemical properties of the NPs, including size, shape, surface area, and chemistry, as they increase reactivity and potential for interactions, stability, and surface charge. Because charged nanoparticles can interact differently with biological systems and reactivity, they significantly influence behavior and interactions within ecosystems. Different NPs have varying chemical compositions, which can affect their toxicity and environmental persistence. Understanding the composition is crucial for assessing potential risks. By their area of use, NPs or the products containing them can be grouped into building/construction, coatings/inks, cleaning/disinfection, complex objects and other products, cosmetics, electronics, food/feed, healthcare, paper products, rubber products, textiles, etc. Hence, to characterize the risk there a need to integrate exposure and toxicity data to quantitatively assess the risks NPs pose to ecosystems. Consider factors such as the likelihood and severity of impacts, as well as the sensitivity and resilience of affected ecosystems.

7.2.2 Bioavailability, Uptake and Toxicity Assessment

Assessing the bioavailability of NPs is essential to understand their uptake and accumulation in organisms. Factors, such as particle size, surface chemistry, and environmental conditions, influence NP bioavailability and the potential for NPs to be taken up by organisms and transferred through food chains. Factors such as surface modification, NP size, and biological barriers should be considered. Due to their size and surface properties, NPs may exhibit increased toxicity compared to their bulk counterparts. Therefore, toxicological studies should be conducted to determine their potential harm to different organisms in the ecosystem. To investigate the effects of NPs on various components of ecosystems, including individual organisms, populations, and communities, may involve laboratory experiments, field studies, and modeling approaches. A number of concepts for the grouping and efficient testing and assessment of nanomaterials (NMs) have been published that are based on NM grouping on their structure and material properties and functions, their exposure, uptake, and kinetics, and initiating cellular or apical effects. There is a predictive toxicological approach [5] using mechanism-based *in-vitro* assays for high-throughput screening (HTS) of NPs, which simultaneously measures sublethal and lethal cellular endpoints reflecting important toxicity pathways for NPs, e.g., inflammatory reactions induced by Ag and ZnO NPs that release toxic ions, surface reactivity-induced redox activity and ROS production by TiO_2 NPs, membrane lysis elicited by partly soluble SiO2 NPs, or fiber-like toxicity caused by high aspect ratio carbon nanotubes.

Ecotoxicity could be "acute toxicity" that can cause immediate harmful effects on organisms. There could be "chronic toxicity" that is long-term effects, such as developmental or reproductive issues. Even if not immediately lethal, nanoparticles may cause sublethal impacts, like behavior changes or impaired reproduction. There is a possibility of "species sensitivity" that varies with different organisms that may react differently to nanoparticles. This will involve use of standardized methods for toxicity testing to ensure consistency.

7.2.3 Exposure Assessment Fate and Transport of NP

Exposure assessment is done by evaluating the pathways through which NPs may enter the ecosystem, such as direct release into water bodies, soil, or air, or indirect pathways through waste streams or bioaccumulation in organisms. Or NPs can move through air to water, and to soil, leading to exposure of various organisms in all these ecosystems. Evaluating how NPs enter ecosystems through different exposure routes (e.g., inhalation, ingestion, dermal contact) is essential for understanding potential risks to organisms. Moreover, considering the potential long-term effects of NP exposure on ecosystems is important for understanding cumulative impacts and ecosystem resilience. How NPs behave once released into the environment includes understanding their dispersal, aggregation, deposition, and persistence in different environmental matrices. Exposure pathways will vary as per the source of NPs. For example, inhalation by organisms is especially relevant for airborne NPs released during production or use. Whereas, when NPs are directly released into water bodies or through wastewater and contaminated soils, the affected plants and soil organisms are bound to be involved in the food chain, which will have potential for bioaccumulation in organisms and movement through the food web [6].

7.2.4 Ecological Effects and Risk Characterization

Nanoparticles can interact with different components of ecosystems, including organisms, soil, water, and sediments. Assessing these interactions is important for understanding potential ecological impacts. Environmental factors, such as pH, temperature, humidity, and presence of other chemicals, can influence the behavior and toxicity of NPs in ecosystems. The environmental fate will depend on how stable the NPs are under different environmental conditions. Do they agglomerate or remain dispersed? Can they be transported long distances by air or water? Do they settle in water bodies or soils? And can they accumulate in organisms through the food chain? By considering these factors, researchers and regulators can develop a comprehensive understanding of the risks posed by NPs to ecosystems and make informed decisions to mitigate those risks.

Specific endpoints related to mortality, growth, behavior, etc., should be measured along with testing on a range of species to understand potential impacts on different parts of the ecosystem. There is a need to have realistic exposure scenarios that mimic real-world exposure conditions. Therefore, to know the ecological effects, evaluate the broader ecological impacts of NP exposure, including effects on ecosystem structure and function, species interactions, biodiversity, and ecosystem services. To assess their effect on the environment, Suchi *et al.* [7] reviewed the assessment of NPs in the environment using the causal diagram approach. It is an integrated approach to comprise available scientific information within an interdisciplinary logical framework, to identify knowledge gaps, and to describe environment and health linkages for natural NPs and engineered NPs. The causal diagram can handle the complexity of issues on NP safety, from their exposure to the effects on the environment and health. It considers available scientific information starting with common sources of NPs and their interactions with various environmental processes that may pose threats to both human health and the ecological risk. This user-friendly tool represents a valuable information package for various stakeholders.

7.2.5 Uncertainty and Variability

According to the United States Environmental Protection Agency (EPA), ***Variability*** *refers to the inherent heterogeneity or diversity of data in an assessment. It is "a quantitative description of the range or spread of a set of values," and is often expressed through statistical metrics such as variance, standard deviation, and interquartile ranges that reflect the variability of the data.* [8]. Variability cannot be reduced, but it can be better characterized.

And ***Uncertainty*** *refers to a lack of data or an incomplete understanding of the context of the risk assessment decision. It can be either qualitative or quantitative* [8]. It can be reduced or eliminated with more or better data. Qualitative uncertainty may be due to a lack of knowledge about the factors that affect exposure, whereas quantitative uncertainty may come from the use of non-precise measurement methods. Chemical concentrations in environmental media can be approximated using assumptions (more uncertainty) or described using measured data (less uncertainty). Uncertainty can be introduced when defining exposure assumptions, identifying individual parameters (i.e., data), making model predictions, or formulating judgments of the risk assessment.

The factors that contribute to variability and uncertainty in exposure assessment also differ. For example, in an exposure assessment, variability may be present in the measurements of source contaminant concentrations;

environmental parameters (e.g., pH, temperature, or other factors that influence fate, transport, and transformation of a contaminant); human exposure factors (e.g., age, gender, behavioral patterns, location, socioeconomic factors, individual susceptibilities). When considering a population's exposure to urban air pollution, variability may exist in the measured pollution exposure concentrations due to when, where, and how the different measurements were taken (e.g., weekday vs. weekend measurements near roads in a city's financial district can be very different; measurements from personal air monitoring devices will vary across individuals due to individuals' behavior, such as how much time they spend outdoors). Variability may also exist in the population itself, which could explain variability in the exposure measurements. For example, members of a study population who bike to work on busy roads may have higher exposure to air pollution than members of the population who commute in a vehicle with the windows closed. Younger members of the study population may have faster breathing rates than older members, resulting in greater exposures. Whereas for uncertainties, the exposure scenario uncertainty might result from (i) descriptive errors; (ii) aggregation errors; (iii) professional judgment errors; (iv) incomplete analysis; (v) measurement or sampling errors; or (vi) model uncertainty (e.g., relationship errors, parameter uncertainty, selection of incorrect model, modeling errors). Uncertainty in risk assessment can be present in the characterization of the exposure scenario, the parameter estimates, and model predictions.

It is important to recognize and address the uncertainties associated with NP risk assessment, including variability in environmental conditions, biological responses, and the extrapolation of laboratory findings to real-world scenarios. While nanoparticles offer many promising applications, their uncertainties and variability highlight the need for careful consideration in their production, use, and disposal. Advances in research, characterization techniques, and regulatory frameworks will be crucial in harnessing the potential of nanoparticles while minimizing risks. Due to their unique properties and behaviors at the nanoscale, NPs present specific challenges and uncertainties when it comes to risk assessment. One of the primary uncertainties is understanding the toxicity of nanoparticles. Their small size and high surface area can lead to interactions with biological systems that are not well understood. The toxicity of nanoparticles can vary significantly based on their size, shape, surface chemistry, and other factors. Other uncertainties related to variability could be routes of exposure, fate of NPs once released into atmosphere, dose response effect, cumulative or long-term effect, modeling challenges and regulatory gaps. Addressing these uncertainties requires interdisciplinary collaboration among scientists, engineers, toxicologists, regulators, and policymakers. Continued research, improved measurement techniques, standardized protocols, and updated regulations are needed.

The OECD/ECHA analogue and category approaches are described as techniques for grouping chemicals. The analogue approach implies that grouping is based on a limited number of chemicals. Endpoint information for one chemical is used to predict the same endpoint for another chemical, which is considered to be similar, even though trends in properties (a term used in the OECD guidance to describe regular patterns of properties as a result of structural (or other) similarities) are not necessarily apparent. Therefore, not every chemical needs to be tested for every required endpoint. The overall category data and rationale should be adequate to support a screening-level hazard assessment [9–11].

7.2.6 Risk Management and Mitigation

Risk management is the process of identifying, analyzing, and addressing risks. Risk mitigation is the process of taking actions to reduce the probability of risks occurring and minimizing their impact. Risk management and mitigation develops strategies to minimize or mitigate the risks associated with NPs, such as pollution prevention measures, regulatory controls, or the development of safer alternative materials. Compliance with existing regulatory frameworks for assessing and managing risks associated with NPs in the environment is crucial. This includes regulations related to chemical safety, environmental protection, and waste management. By considering these factors in NP risk assessments, researchers and policymakers can better understand the potential impacts of NPs on ecosystems and develop appropriate mitigation strategies to protect environmental health. To evaluate hazard information, the dose descriptor of the substance as per DNEL (derived no-effect level) has to be identified, which represents a level of exposure above which humans should not be exposed. Moreover, the PNEC (predicted no-effect-concentration), which is the concentration of a chemical in any compartment below which unacceptable effects on ecosystem and its organisms will most likely not occur during long-term or short-term exposure, also has to be considered. Evidence from industrial case studies on the impact and effectiveness of risk mitigation strategies on the insurability of nanomaterial production was presented by Bergamaschi *et al.* [12]. Using physicochemical data provided by various sources, the results of the hazard identification were compiled into a hazard matrix table to see the role of remediation strategies on the intrinsic hazard potential of the different nanomaterials. The demonstrated possibility to tune hazardous properties and exposure determinant allows engineered NP production risk to reside within acceptability, and a shift toward lower exposure and hazard bands categories should be implemented and pursued as a best practice and proactive behavior against the simple precautionary approach.

Several organizations and regulatory agencies have developed guidance documents and frameworks for assessing the risks associated with NPs. Some of these include:

(i) *Organization for Economic Cooperation and Development (OECD):* Several guidance documents related to the risk assessment of nanomaterials have been developed by the OECD, including:
- Guidance Manual for the Testing of Manufactured Nanomaterials (Series on the Safety of Manufactured Nanomaterials)
- Guidance on Sample Preparation and Dosimetry for the Safety Testing of Manufactured Nanomaterials
- Guidance on Grouping of Chemicals
- Guidance on Aquatic and Sediment Toxicological Testing of Nanomaterials.

(ii) *European Commission (EC):* The EC has published the following guidance documents and reports on nanomaterial risk assessment:
- Guidance on the Regulation of Nanomaterials
- Guidance on Information Requirements and Chemical Safety Assessment - Chapter R.6: QSARs and Grouping of Chemicals
- Nanomaterials: Impact on Human Health
- Nanomaterials: Impact on Environment
- OECD Test Guidelines for the Testing of Chemicals – Section 2: Effects on Biotic Systems – Test No. 318: Dispersion Stability of Nanomaterials in Simulated Environmental Media.

(iii) *United States Environmental Protection Agency (EPA):* The EPA has issued various documents and resources related to nanomaterial risk assessment, such as:

- NanoGrid: A Framework for Evaluating the Environmental and Human Health Impacts of Nanomaterials
- Nanomaterials Research Strategy
- Overview of EPA's Nanomaterials Research Strategy.

(iv) *National Institute for Occupational Safety and Health (NIOSH):* The guidance documents for assessing the occupational health risks of NP produced by NIOSH are:
- Approaches to Safe Nanotechnology: Managing the Health and Safety Concerns Associated with Engineered Nanomaterials
- Occupational Exposure to Titanium Dioxide
- Current Intelligence Bulletin 65: Occupational Exposure to Carbon Nanotubes and Nanofibers.

(v) *International Organization for Standardization (ISO):* The ISO has developed standards related to nanotechnology and risk assessment, such as:
- ISO/TS 80004-2:2015 - Nanotechnologies -- Vocabulary -- Part 2: Nano-objects
- ISO/TR 13121:2011 - Nanotechnologies -- Nanomaterial risk evaluation
- ISO/TS 12901-2:2014 - Nanotechnologies -- Occupational risk management applied to engineered nanomaterials.

(vi) *Government of India (Department of Biotechnology and Ministry of Science & Technology)* (2020)

The Government of India has prepared "Guidelines for Evaluation of Nano-Based Agri-Input and Food Products in India," which includes Nanofertilizer (with or without nano or non-nano carriers); Nanopesticides (with or without carrier, their chemistry, bio-efficacy and residues, toxicity, packaging and processing of Nanopesticides products for manufacture or import) also Nanofood and Nanofeed.

All these documents provide valuable guidance on methodologies, testing procedures, and risk assessment frameworks for evaluating the potential risks associated with NPs to human health and the environment. In summary, here are some key considerations for nanoparticle-related risk assessment in ecosystems:

(i) Consideration of Interactions
- *Synergistic Effects:* Interactions between nanoparticles and other pollutants.
- *Ecosystem Effects:* Impact of changes at one level (e.g., population) to the entire ecosystem.
- *Adaptation:* Possibility of organisms adapting to the presence of nanoparticles, and their consequences.

(ii) Data Gaps and Uncertainties
- *Lack of Data:* Incomplete information on nanoparticle behavior or toxicity.
- *Modeling:* Use of models to predict behavior in the absence of empirical data.
- *Long-Term Effects:* Challenges in predicting long-term impacts on ecosystems.

(iii) Communication and Stakeholder Involvement
- *Transparency:* Clear communication of risks and uncertainties to stakeholders.
- *Public Input:* Involvement of communities potentially affected by nanoparticle use.

(iv) Case Studies and Research
- *Existing Studies:* Learn from previous studies on similar nanoparticles.
- *Field Studies:* Actual observations in natural environments can provide valuable insights.
- *Multidisciplinary Approach:* Involvement of experts from various fields (biology, chemistry, ecology, etc.).

7.3 Monitoring and Surveillance

Nanoparticle monitoring and tracking in ecosystems require a multi-faceted approach, integrating various technologies and strategies. Interdisciplinary collaboration between ecologists, chemists, engineers, and data scientists is crucial for effective monitoring, risk assessment, and mitigation of potential ecological impacts.

7.3.1 Monitoring Strategies

Gallego-Urrea *et al.* [13] have explored the feasibility of a novel minimum perturbation method that relates the diffusive movement of NPs in a light field to their size distributions. Their study was based on NP tracking analysis (NTA) of aquatic environmental samples that involved measurements of NP number, concentrations, and size distributions in contrasting aquatic environments. Following are some strategies and technologies that can be employed for NP monitoring and tracking to assess their risks to ecosystems.

7.3.1.1 Remote Sensing

Remote sensing can be done using

(i) *Satellite Imagery:* For larger-scale monitoring, satellite imagery can help detect changes in ecosystems that might be linked to nanoparticle presence.
(ii) *Aerial Drones:* When equipped with sensors, aerial drones can be used to monitor specific areas more closely.
(iii) *In-Situ Monitoring:* Regular sampling of water bodies, air and soil where nanoparticles are suspected to be present. While monitoring of engineered NPs in soil-plant system, the problem faced is in preparation of soil and plant sample, because the structure and composition of NPs may change in response to their environment; hence, results obtained for processed or digested samples can often differ from a situation where the particles are characterized *in situ* as well as in their pristine state. Therefore, as suggested by Shrivastava *et al.* [14], analysis should be performed by using techniques that require minimum sample preparation such as using sensor and microfludic chips for monitoring ENPs in soil-plant system.
(iv) *Deploying Sensor Networks:* Sensor networks can be deployed for continuously monitoring nanoparticle levels.
(v) *Biological Monitoring:* Bioindicators can be achieved using specific species as indicators of nanoparticle presence and effects; for example, changes in fish behavior or growth rates and other animal models [15]. Monitoring can also be done by biomagnification studies, i.e., monitoring how NPs move through food chains.

7.3.1.2 *Tracking Technologies*

Monitoring programs are implemented to track the presence and behavior of NPs in the environment over time, as well as any associated ecological changes or emerging risks that, must be addressed This is done through different tracking technologies such as

(i) *Nanotechnology-Based Tracers:* Fluorescent NPs can be designed to mimic the behavior of other nanoparticles and can be tracked using fluorescence detectors. Or radioactive NP tracers can be made radioactive (with safe isotopes) for tracking using radiation detectors.
(ii) *Mass Spectrometry:* Either ICP-MS (inductively coupled plasma mass spectrometry): technique can detect and quantify NPs based on their mass or NanoSIMS (nano-scale secondary ion mass spectrometry) can be used, which provides very high spatial resolution. Both are useful for tracking nanoparticles in specific tissues or organisms.
(iii) *Chemical Tags:* NPs can be surface modified with unique chemical tags that can be detected later. Isotopes can also be used as labels for tracking.
(iv) *Machine Learning and Data Analysis:* With advancement in AI, there is a possibility to utilize AI and machine learning algorithms to analyze monitoring data that can provide insights into nanoparticle behavior and risks.
(v) *Real-Time Monitoring:* This can be done by developing smart sensors that can detect and quantify nanoparticles in real time. The IoT (internet of things) also creates interconnected systems for continuous monitoring and data collection.

Guo *et al.* [16] realized that the methods capable of controlling synthesis at the level of an individual nanoparticle are needed for reproducibility and scalability in engineering complex nanomaterials. So, they combined the spatially patterned activation of the photo-reductant sodium pyruvate with interferometric scattering microscopy to achieve fast, label-free monitoring and control of hundreds of gold NPs in real time. They established a combination of label-free iSCAT microscopy and photo-reductant spatiotemporal control as a simple means of studying NP kinetics in real time at the level of individual particles in a manner that is independent of the optical properties of the NP.

7.3.1.3 Challenges and Considerations

Challenges that can be envisaged are in developing standardized methods for NP monitoring and tracking. Interpretation or understanding the ecological implications of presence of NP requires interdisciplinary collaboration. Many of these technologies can be expensive to implement. Above all, proper regulatory-frameworks are of utmost importance. Developing regulations around NP use and monitoring is crucial for ecosystem protection.

The EU NanoDefine project requires case studies and research focused on developing methods for the detection, identification, and quantification of engineered nanoparticles. The tracking of NP in aquatic environments is a big challenge which is being taken up by research projects similar to "nanoprobes," which aim to understand NP behavior in marine environments.

There have been studies focused on challenges faced in monitoring specific NPs by different groups, e.g., for nano-TiO_2 [17] and for nanosilver [18].

7.4 Stakeholder Engagement

Assessing risks to ecosystems posed by nanoparticles involves a wide range of stakeholders, each with their own interests, responsibilities, and expertise. Effective communication and collaboration among all the stakeholders are crucial. This interdisciplinary approach helps to ensure that scientific understanding, regulatory frameworks, industry practices, and public concerns are all considered in a comprehensive risk assessment process. Following are some of the key stakeholders typically involved in assessing these risks.

7.4.1 Government Agencies

Various government bodies at the local, national, and international levels are crucial stakeholders. These may include environmental protection agencies, health departments, and regulatory bodies. They are responsible for developing regulations, policies, and guidelines related to the use of nanoparticles and assessing their potential risks to ecosystems. According to a 2017 study, government agencies are one of several stakeholders that have opinions about the risks of nanomaterials and how to manage them [19].

7.4.2 International Organizations

International organizations such as the United Nations Environment Programme (UNEP) and the World Health Organization (WHO) may also be stakeholders. They work on developing international guidelines and standards for nanoparticle use and risk assessment. International organizations could take responsibility for developing and promoting a nano-specific international legal framework to protect the environment from nanoparticles and their risks. This new legal framework could incorporate the precautionary principle, technology transfer, research cooperation, and a duty to promote transparency.

7.4.3 Researchers and Scientists

Researchers and scientists from various fields, such as environmental science, toxicology, chemistry, and materials science, are essential stakeholders. They conduct studies to understand the behavior, fate, and impacts of nanoparticles in ecosystems. Their research forms the basis for risk assessments. Academic institutions, universities, and research institutions contribute significantly to the understanding of NP risks. They conduct research, educate and train future scientists and policymakers, and often collaborate with government agencies and industry partners.

7.4.4 Health Professionals

Doctors, nurses, and public health officials are stakeholders due to their roles in understanding the health impacts of NPs, both on ecosystems and human health. Their input is valuable for assessing risks comprehensively.

7.4.5 Industry

Companies that produce, use, or dispose of NPs are important stakeholders. They have a vested interest in understanding and managing risks associated with their products. Industry stakeholders often collaborate with researchers and regulatory agencies to ensure compliance with regulations and to develop safe practices. Moreover, supplier companies of raw materials for NP production may also have a stake in ensuring their products are used responsibly and safely.

7.4.6 Non-Governmental Organizations (NGOs)

Many NGOs related to environmental and consumer advocacy groups play a role in advocating for responsible use of nanoparticles. They often conduct their own research, raise awareness, and provide input during policy development.

7.4.7 Consumers

Consumers and public interest groups who use products containing NPs or are potentially affected by their presence in the environment are also stakeholders. Their concerns and input are important in shaping regulations and policies.

7.4.8 Media

Last but not the least, the media and the public play a role in disseminating information about NP-related risks, influencing public perception, and thereby influencing the actions of other stakeholders. The public's understanding and concerns can also drive policy and regulatory decisions.

It is important to engage stakeholders, including scientists, policymakers, industry representatives, and the public, throughout the risk assessment process to ensure transparency, accountability, and informed decision-making. Moreover, it is necessary to develop strategies to communicate risks effectively to different stakeholders. By following these steps and incorporating the latest scientific knowledge and methodologies, researchers and policymakers can better understand and address the potential risks NPs pose to ecosystems.

7.5 Concluding Remarks

Assessing the risks of NPs to ecosystems involves considering various areas to ensure a comprehensive evaluation. The unique properties of NPs may pose risks to the environment. The key areas that need assessing include direct toxicity of the nanoparticles to different species to determine if NPs can accumulate in organisms over time, potentially leading to biomagnification in food chains. For this it is important to understand how NPs are distributed within organisms and ecosystems, which will help in knowing the ecological impact.

Ecosystem disruption leads to effects on biodiversity and habitat alteration. It is necessary to evaluate if NPs cause changes to habitats that affect the survival of species. Some organisms may benefit from nanoparticles, leading to imbalances in ecosystems. This depends on how easily NPs can move through different environmental compartments (soil, water, air). Other important criteria are: (i) Persistence (i.e., determine how long nanoparticles remain in the environment before degrading), (ii) Transformation (i.e., whether NPs change form or properties as they move through the environment), (iii) Exposure Pathways (by evaluating routes of exposure for organisms by ingestion, inhalation, dermal contact, etc.), or (iv) Indirect/Secondary Exposure Pathways, such as bioaccumulation through the food chain. These demand monitoring, detection and quantification of NP in environmental samples, along with monitoring programs to track NP levels in ecosystems over time. Nano-QSAR in the context of agricultural nanomaterials has already been discussed in the previous chapter (6.6.3).

Regulatory and Policy makers have to identify the gaps in existing regulations related to NP risks and develop strategies to mitigate risks, such as guidelines for safe disposal or use.

Interacting stressors like synergistic effects of NP with other pollutants or stressors in the environment; and climate change need assessment. There could be potential interactions between NP exposure and effects of climate change on ecosystems.

All these considerations demand thorough in-depth research to know the data gaps that will help in identifying areas where more research is needed to understand NP-related risks fully. This should cover long-term effects of NP exposure on ecosystems along with case studies and previous incidents where NPs caused harm to ecosystems. Finally, best practices are to be identified for handling and using NPs to minimize risks.

In conclusion, it can be said that assessing the risks of NPs to ecosystems is a complex and interdisciplinary task. It requires expertise in toxicology, ecology, chemistry, and environmental science. By considering these areas, researchers and policymakers can develop a comprehensive understanding of the potential impacts of NPs and work towards sustainable and responsible use in our environment.

References

1. Kumah, E.A., Fopa, R.D., Harati, S., *et al.*, Human and environmental impacts of nanoparticles: a scoping review of the current literature. *BMC Public Health*, 23, 1059, 2023, https://doi.org/10.1186/s12889-023-15958-4.

2. Yin, X.-H., Xu, Y.-M., Lau, A.T.Y., Nanoparticles: Excellent materials yet dangerous when they become airborne. *Toxics*, 10, 50, 2022, https://doi.org/10.3390/toxics10020050.
3. Graeme, E.B., Jason, K.K., Maclaughlin, M.J., Fate and risks of nanomaterials in aquatic and terrestrial environments. *Acc. Chem. Res.*, 46, 3, 854–862, 2013.
4. Ren, W., Liu, H., Teng, Y., Impact of nanoparticles on soil ecosystems, in: *The Role of Nanoparticles in Plant Nutrition under Soil Pollution*, 2022.
5. Nel, A.E., Xia, T., Meng, H., Wang, X., Lin, S., Ji, Z., Zhang, H., Nanomaterial toxicity testing in the 21st century: use of a predictive toxicological approach and high-throughput screening. *Acc. Chem. Res.*, 46, 607–621, 2013.
6. Kuempel, E.D., Geraci, C.L., Schulte, P.A., Risk assessment and risk management of nanomaterials in the workplace: Translating research to practice. *Ann. Occup. Hyg.*, 56, 5, 491–505, 2012.
7. Suchi, S., Gupta, S.K., Bartonova, A., Dusinska, M., Gutleb, A.C., Rahman, Q., Nanoparticles in the environment: Assessment using the causal diagram approach. *Environ. Health*, 11, Suppl 1, S13, 2012. http://www.ehjournal.net/content/11/S1/S13.
8. U.S. EPA (U.S. Environmental Protection Agency), *Exposure Factors Handbook 2011 Edition*, 2011, (Final) [EPA Report]. (EPA/600/R-09/052F).
9. Arts, J.H.E., Hadi, M., Keene, A.M., Kreiling, R., Lyon, D., Maier, M., Michel, K., Petry, T., Sauer, U.G., Warheit, D., Wiench, K., Landsiedel, R., A critical appraisal of existing concepts for the grouping of nanomaterials. *Regul. Toxicol. Pharm.*, 70, 492–506, 2014.
10. OECD, Guidance on grouping of chemicals. Series on testing and assessment No. 194 (second ed.), ENV/JM/MONO(2014) 4, OECD, Paris, France (14.04.14), 2014.
11. ECHA, Guidance on information requirements and chemical safety assessment. Chapter R.6. QSARs and grouping of chemicals. *European Chem. Agency,* p. 134. Available at: http://echa.europa.eu/documents/10162/13632/information_requirements_r6_en.pdf, 2008.
12. Bergamaschi, E., Murphy, F., Poland, C.A., Mullins, M., Costa, A.L., McAlea, E., Tran, L., Tofail, S.A., Impact and effectiveness of risk mitigation strategies on the insurability of nanomaterial production: evidences from industrial case studies. *Wiley Interdiscip. Rev. Nanomed. Nanobiotechnol.*, 7, 6, 839–855, 2015, doi: 10.1002/wnan.1340.
13. Gallego-Urrea, J.A., Tuoriniemi, J., Pallander, T., Hassellöv, M., Measurements of nanoparticle number concentrations and size distributions in contrasting aquatic environments using nanoparticle tracking analysis. *Environ. Chem.*, 7, 1, 67–81, 2010, https://doi.org/10.1071/EN09114.

14. Shrivastava, M., Srivastav, A., Gandhi, S., Rao, S., Roychoudhury, A., Kumar, A., Singhal, R.K., Jha, S.K., Singh, S.D., Monitoring of engineered nanoparticles in soil-plant system: A review. *Environ. Nanotechnol. Monit. Manage.*, 11, 100218, 2019. ISSN 2215-1532, https://doi.org/10.1016/j.enmm.2019.100218.
15. Cho, E.J., Holback, H., Liu, K.C., Abouelmagd, S.A., Park, J., Yeo, Y., Nanoparticle characterization: state of the art, challenges, and emerging technologies. *Mol. Pharm.*, 10, 6, 2093–110, 2013, doi: 10.1021/mp300697h. Epub 2013 Mar 21. PMID: 23461379; PMCID: PMC3672343.
16. Guo, Y., Walter, V., Vanuytsel, S., Parperis, C., Sengel, J.T., Weatherill, E.E., Wallace, M.I., Real-time monitoring and control of nanoparticle formation. *J. Am. Chem. Soc.*, 145, 29, 15809–15815, 2023, https://doi.org/10.1021/jacs.3c02484.
17. Heilgeist, S., Sekine, R., Sahin, O., Stewart, R.A., Finding Nano: Challenges involved in monitoring the presence and fate of engineered titanium dioxide nanoparticles in aquatic environments. *Water*, 13, 5, 734, 2021, https://doi.org/10.3390/w13050734.
18. Stensberg, M.C., Wei, Q., McLamore, E.S., Porterfield, D.M., Wei, A., Sepúlveda, M.S., Toxicological studies on silver nanoparticles: challenges and opportunities in assessment, monitoring and imaging. *Nanomed. (Lond)*, 6, 5, 879–98, 2011. doi: 10.2217/nnm.11.78. PMID: 21793678; PMCID: PMC3359871.
19. Capon, A., Gillespie, J., Rolfe, M., Smith, W., Perceptions of risk from nanotechnologies and trust in stakeholders: a cross sectional study of public, academic, government and business attitudes. *BMC Public Health*, 15, 424, 2015, doi: 10.1186/s12889- 015-1795-1.

Part II
ETHICAL ASPECTS

8

Ethics Governance and Policies: Problems and Concerns

Ethics and Science need to shake hands.

Richard Cabot

Abstract
This chapter introduces the ethical dimensions of nanotechnology. It emphasizes the importance of ethical governance and policy frameworks from research labs to consumer markets. Ethical concerns include privacy, security, environmental impact, and organizational responsibility. Challenges in implementing ethics include subjectivity, complexity, limited awareness, resource constraints, and legal compliance. Given the irreversible and long-term impacts of nanotechnology, ethical decision-making must anticipate future scenarios despite uncertainties. This chapter outlines foundational elements for creating ethics-based governance models that support responsible innovation in nanotech.

Keywords: Ethical-framework, policies, governance, security, equity, cultural-sensitivity, ethical-concerns, legal-regulation

8.1 Introduction

The definition of "ethics" as per the *Oxford Dictionary* is: "*Moral principles that govern a person's behavior or the conducting of an activity.*" Ethics, which refers to the philosophical study of moral principles, is distinct from religion, which may inform ethical beliefs but is not synonymous with it. But ethics in science do enter into the question. Ethics examines the rational justification for our moral judgments; it studies what is morally right or wrong, just, or unjust. In a broader sense, ethics reflects on human beings and their interaction with nature and with other humans, on freedom, on responsibility and on justice. Ethical value involves considerations such as

Madhuri Sharon. Risk, Safety, and Ethical Concerns in Nanotechnology, (165–186) © 2025 Scrivener Publishing LLC

choice, conscience, fairness, honesty, honor, integrity, moral, principles, responsibility, and rights.

An ethical issue is a circumstance in which a moral conflict arises in the workplace; thus, it is a situation in which a moral standard is being challenged. Ethical issues in the workplace occur when a moral dilemma emerges and must be resolved within a corporation.

Ethics is based on socio-cultural or religious beliefs of what is righteous and what is "immoral" or "evil." Ethics encompasses respect for the community, justice, rights, human dignity, and the autonomy of the individual. For science and technology, that affects the environment and health policy; ethics is concerned with organizing financing and delivering the product. In this respect, ethics is a bridge between healthy policy and values, where the moral validity of the choice is examined.

Science and technology are directly related to society as they can induce societal changes. Scientists must very responsibly consider the ethical and social aspects of their work. Ethics in science and technology are the values and ethical considerations that guide the regulation and oversight of technology, protect privacy, and require responsible actions in its creation and use. Ethics in science refers to moral principles that guide an individual or organization in their scientific pursuits, and is concerned with what is right and wrong. This includes honesty, integrity, and fairness. It also includes the values, social responsibilities, and codes of conduct that should be held in relation to scientific and technological innovations. Where human subjects are involved, science and technological research must be informed about their disease or treatment for it, even after cures are discovered. Ethical rules in science include honest reporting, avoiding bias, and treating subjects humanely. Ethics promote the aims of scientific research, including knowledge, truth, and avoidance of error. Any deviation from ethics in scientific research may result in research misconduct, including falsification, fabrications, and plagiarism.

Technology is the outcome of scientific innovation, which progresses with the use if equipment, apparatus, machine, hardware, mechanism, and gadgets controlled by automation and industry. Technology demands adhering to ethics more strictly than science, because technology is the application of scientific knowledge for practical purposes that reaches the masses through the involvement of industry. It is heartening to realize that significant considerations are made on the close connection between ethics and science and technology; and their effects on society, the environment, and values.

This chapter concentrates on the risk aspect of nanotechnology, which involves working at the atomic, molecular, and supramolecular levels limited to a size scale of 1–100 nm. The use of nano-size materials has found application in several areas because they exhibit special structures, new properties, and functions. Nano-sized devices have application in areas such as industrial bioprocessing, molecular medicine, personalized medicine; medical and environmental diagnostics, food processing, packaging, agricultural systems, electronics, telecommunications, defense, weapons, and the list goes on. This raises the question of ethical issues at all the stages where nanoparticles are involved—from research, production, use to disposal.

8.2 How Does Ethics Pertain to Nanotechnology?

More than three decades ago, an ethical concern regarding the harmful impact of nano-size and larger size particles when inhaled was raised in a study [1]. The study included nanoparticles of titanium dioxide (TiO_2) and aluminum oxide (Al_2O_3). Increased phagocytotic uptake of nanoparticles was found to be more harmful, causing a bigger risk than larger size particles. Ethical concerns were raised regarding the acceptability, sustainability, and societal desirability of the nano-based marketable products [2]; and a challenge for an ethical foundation for considering the future consequences of their actions and to make decisions early on to avoid significant liability risk was taken up by many international organizations, such as the European Commission, which made recommendations on a "Code of Conduct for Responsible Nanoscience and Nanotechnology Research" [3].

Like any advancing technology, nanotechnology presents a range of ethical, governance, and policy challenges. More so, because it is being used for many different types of applications and affects not only human beings but also the environmental ecosystem, including animals, plants, and microorganisms of both terrestrial and aquatic domain. It is crucial to understand how engineered nanoparticles interact with ecosystems. Ethical questions revolve around the impact of these particles on soil, water, and wildlife. As far as the environment is concerned, released nanoparticles can find their way into different ecosystems through various means such as industrial waste, consumer products, and even byproducts of medical applications. The long-term impact of these particles on ecosystems and human health is not yet fully understood. Nanoparticles may have different toxicological effects compared to their larger counterparts. This raises concerns about their varied impact on plants, animals, and microorganisms in the environment. Ethical considerations should include how to minimize exposure during manufacturing, use, and disposal.

Current regulations may not be sufficient to manage the potential risks of engineered nanomaterials. There is a need for updated policies and frameworks to ensure their safe use and disposal. Health and safety of workers as well as consumers both are at risk. Those working with nanomaterials in research labs or industrial settings face potential health risks due to direct exposure. Proper safety protocols and regulations are necessary to protect these workers. Moreover, nanoparticles are increasingly used in consumer products like cosmetics, clothing, and food packaging. Questions arise about whether these products are adequately tested for safety and whether consumers are aware of the presence of nanoparticles. However, the option of using nanoscale sensors to test consumer products raises concerns about pervasive surveillance, since tiny sensors could potentially be used for constant monitoring, infringing on privacy rights.

Other areas that need addressing are equity and access, since there is a risk that nanotechnology could widen the gap between developed and developing nations. Access to benefits such as improved healthcare (like nanomedicine) might not be equitable. And since nanotechnology research requires significant resources, concerns exist about whether these resources should be directed towards more immediate societal needs, like basic healthcare and education.

The race to patent nanotechnologies raises questions about access and affordability. Will the benefits of nanotechnology be restricted to those who can afford patented products and treatments? Balancing the protection of intellectual property with the need for open access to knowledge and technologies is an ongoing challenge. The defense sector and military is spending heavily on nanotechnology, causing concern for their ethical use. Nanotechnology has various military applications, from improving armor to creating more advanced weaponry. The ethical implications of these developments, such as increased lethality or potential for misuse, are significant. Although nanotechnology holds promise for enhancing human capabilities, such as cognitive function or physical abilities, ethical questions arise about who should have access to these enhancements and whether they could create social divides. Moreover, nanotechnology inventions and discoveries raise questions about who owns and controls these technologies. Ethical considerations involve balancing the benefits of innovation with fair access and distribution. Ethical guidelines have to ensure fair competition and access to patented technologies, so as to avoid monopolies and restrict access.

Like any new technology, nanotechnology can have unforeseen unintended consequences. Ethical considerations must encompass the potential unintended outcomes of nanotech applications. Who should be held responsible for any negative effects of nanotechnology? Establishing accountability frameworks is crucial. Addressing these concerns require a multi-stakeholder approach involving scientists, policymakers, industry leaders, ethicists, and the public. Robust ethical frameworks, updated regulations, transparency in research, and responsible innovation are essential for harnessing the benefits of nanotechnology while minimizing its risks.

Here are some key responsibilities and concerns associated with nanotechnology that need to be addressed by governance and regulation at the international collaboration level because nanotechnology is a global endeavor, requiring international cooperation for effective regulation. Harmonizing standards and regulations across different countries is a challenge, and regulations often struggle to keep pace with the rapid development of nanotechnology. This can lead to gaps in oversight and potential risks going unaddressed. Furthermore, there is a need for transparency in research and development processes involving nanotechnology. This includes disclosing potential risks and involving the public in decision-making. The ethics pertaining to nanotechnology involves the following moral principles, guidelines, and considerations that apply to the development, deployment, and use of nanomaterial:

Ecological or environmental consequences pertains to the following area:

(i) *Health and Safety:* This is also needed at the research level. Because there has to be informed consent when human subjects are involved in nanotechnology research, obtaining informed consent is essential. This includes informing participants about potential risks associated with nanomaterial exposure.

(ii) *Nanotechnology Surveillance*: The use of nanoscale sensors and devices raises concerns about privacy infringement. Ethical considerations involve respecting the privacy of individuals while using these technologies for surveillance or data collection.

(iii) *Security Risks:* Because nanotechnology can lead to the development of powerful surveillance tools or new forms of weaponry, security risks come into the picture. Ethical guidelines are needed to prevent misuse and protect against these threats.

(iv) *Equity and Access:* The technological divide poses a significant risk that nanotechnology advancements could widen the gap between wealthy and poorer nations or communities. Ethical considerations involve ensuring equitable access to benefits and addressing potential disparities. Moreover, healthcare disparities are a concern. Since nanotechnology offers promising advances in medicine, ethical frameworks should address how to avoid unequal access to the treatments provided by these innovations to ensure they are accessible to all.

(v) *Ethical Research Practices:* Transparency and openness about research methods, findings, and potential risks is crucial. Ethical guidelines should emphasize transparency in nanotechnology research. Innovation should be responsible, using the precautionary principle, which suggests that in cases of scientific uncertainty, actions should err on the side of caution to prevent harm. Ethical frameworks may advocate for applying this principle in nanotechnology development. For this societal engagement: involving stakeholders, including the public, in decision-making about nanotechnology can lead to more responsible and ethically sound outcomes.

(vi) *Cultural Sensitivity and Ethical Values:* Nanotechnology applications should respect diverse cultural norms and values. Ethical considerations involve understanding how different cultures perceive and interact with nanotechnology. Accordingly, ethical codes specific to nanotechnology should be developed and adhered to in order to guide researchers, policymakers, and businesses in making morally sound decisions.

(vii) *Addressing Ethical Considerations:* In addressing ethical considerations, organizations, researchers, policymakers, and the public must engage in ongoing dialogue and collaboration. Ethical frameworks can help navigate the complexities of nanotechnology, ensuring its development and use align with societal values and priorities. When developing nanotechnology, how much progress should be allowed and the limit of admissible implication should be considered. There is an urgent need to consider the legal, ethical, and social implications. A question was raised some time ago about the standard view of an ethics for nanotechnology; as many feel that none of the current trends in the discipline of ethics would qualify for application to nanotechnology [4].

8.2.1 Ethics and Nanotechnology in the Workplace

Workplace ethics for nanotechnology-related activities is the first concern, as they are potentially harmful to the health of the workers or researchers due to their exposure to the nanoparticles. Generally, harmful effects are identified after a long period of activity. The discoveries and developments of different types of nanoparticles in the 1990s and early 2000s led to an explosion of interest in nanotechnology. Since then, nanotechnology has continued to advance rapidly, with applications in materials science, electronics, medicine, agriculture, and many other fields. Hence, it has been nearly three and half decades since researchers and workers across the globe have been getting exposed to nanoparticles. This a reasonably long duration for collecting data on the impact of nanoparticles on human health and environment. The world should take lessons from history and not delay in assessing the harmful effects of nanoparticles, as it was late in noticing the harmful effects of radiation caused by the atomic energy used for war. The same may not be exactly true for nanotechnology; however, addressing concerns about the risks involved in its use should not be delayed. Now there are many international and national bodies, as well as researchers who are looking at the possible dark side of nanotechnology [5, 6]. It is necessary to have knowledge about the potentially negative health effects of workplace exposure to nanoparticles so that proper guidelines for decision-making regarding risks and controls can be established and so that decision-makers, employers, workers, investors, and health authorities become aware of these issues. This will help in controlling risks while keeping ethics in mind. To implement the objective of occupational safety and health and the prevention of accidents and diseases in workers, it is imperative that workers should have awareness and acceptance and authorities should implement strict control. Scientists, authorities, and employers must identify and convey the hazards and risks. Moreover, there should be investment in medical screening and toxicological research [13].

Ethical issues involved in nanotechnology in the workplace should respect and allow privacy so that independent and impartial determination of risks, implacability, autonomy, justice, and preservation of privacy are promoted. Discussion on the workplace hazards of nanotechnology should be reinforced in small businesses with the adoption of a global perspective. To implement ethical practices on nanotechnology in the workplace, a state of knowledge about the dangers of nanomaterials and the hazards and risks to workers' health is required, which is still incipient. So, the ethical practices can be implemented only after a provisional assessment of the hazards and risks is carried out for particular nanoparticles. Then, related ethics can be applied in accordance with principles of autonomy, beneficence, nonmaleficence, justice, privacy, and respect for persons.

8.3 Framework for Ethical Analysis of Nanotechnology

In the wake of social reactions to biotechnology; technological advancements have always been challenged, be it genetically manipulated organism (GMO) products; toxicity concerns of use of natural versus synthetic chemicals [7], or the use of synthetic chemicals in food packaging [8]. But the difference between GMOs and nanotechnologies resides in the fact that nanotechnology has so far not attempted to transform the activity and nature of human or living beings. The framework for ethical analysis in nanotechnology is a comprehensive approach that aims to ensure that nanotechnological advancements are developed and deployed in a responsible and ethical manner, considering the wide range of potential impacts on society, the environment, and individuals, especially their health.

8.3.1 Considerations for Framework

(i) *Ethical Principles:* There is a need to review ethical principles in healthcare to ensure that human rights and dignity are fully respected. Promotion of human rights was identified as a core principle of the UN charter, and adopted way back in 1948 in the Universal Declaration of Human Rights; which stated that all people are "free and equal in dignity and rights." Since then, several international treaties, declarations and conventions have been followed. Applying ethical principles such as beneficence (doing good), nonmaleficence (doing no harm), justice (fairness in distribution), autonomy (respecting individuals' choices), and sustainability (considering long-term impacts) to the analysis. No country has the right to set universal standards that other countries must comply with. But there is an irreducible set of international ethical standards common to all societies which have to be respected by all.

(ii) *Identification of Impacts on Specific Issues:* The first step is to identify the specific ethical issues that are relevant to the development and deployment of nanotechnology. For this there is a need to identify the technological source that could have an impact on a specific issue. Here, the dose and duration both become important. Here, risks and impacts are both the same; as they affect many dimensions such as health, mortality, environment, economy, the status and development of scientific research, freedom of choice, privacy, local and national cohabitation; or, as the case may be, international cohabitation, and above all, the more difficult to handle cultural representations of the human being. Hence, these could include concerns about privacy, environmental impact, health and safety risks, economic implications, and more.

(iii) *Continuous Monitoring and Adaptation:* The ethical analysis framework is not static. It should involve mechanisms for continuous monitoring of developments in nanotechnology, reassessment of risks and benefits, and adaptation of ethical guidelines and regulations as needed. Continuous monitoring will help in determining the positive or negative impacts of a specific issue, provided that impact analyses is done by keeping in mind the distinction between scientific uncertainty and scientific ignorance as well as values selected; especially for a comparatively new technology.

(iv) *Stakeholder Analysis:* Understanding who the stakeholders are is crucial. Stakeholders could be researchers, policymakers, industry leaders, consumers, environmental groups, and affected communities. Each stakeholder group may have different perspectives and interests.

(v) *Public Engagement:* Ethical analysis often involves engaging with the public to understand their values, concerns, and preferences. Public perception can influence the development and acceptance of nanotechnology. Decision-makers play a decisive role in adopting ethical principles in healthcare that ensure equity, justice, and respect for human beings. It is also the responsibility of professionals, academicians, as well as ethical human rights and consumer protection groups to supply decision-makers with valid, reliable data so as to ensures equity and justice.

(vi) *Risk Assessment:* Nanotechnology introduces new materials and processes which may have unknown risks.

An ethical framework involves assessing these risks thoroughly. This includes potential risks to human health, the environment, privacy, and social structures. While assessing risk, the quality of health for each person has to be considered, including age or metabolic diseases like diabetes, etc. A consensus should be reached on the values associated with the general issues that relate more specifically to the quality of what is being targeted.

(vii) *Benefit Assessment:* In addition to risks, it's essential to assess the potential benefits of nanotechnology. This includes advancements in medicine, materials science, energy efficiency, and more. Evaluating whether the benefits outweigh the risks is a critical ethical consideration. Moreover, we have to learn from the decisions taken for GMO products, and go for responsible research and innovation as well as follow a shift from a free-market development approach to a responsible development so that necessary focus is brought to take into account the responsibility of all the players involved.

(viii) *Legal and Regulatory Analysis:* Understanding the existing legal and regulatory frameworks is important. This involves looking at whether current laws are sufficient to address the unique challenges of nanotechnology or if new regulations are needed.

(ix) *Decision-Making:* A framework based on risk analysis, benefits, stakeholders' perspectives, and ethical principles helps in making informed decisions. These decisions might involve how to regulate nanotechnology, what research to prioritize, how to communicate risks to the public, and more.

(x) *Equity:* It is often understood as a fundamental expression of social justice. A key aspect of an ethical framework is ensuring transparency in the decision-making processes and holding stakeholders accountable for their actions. Hence, it fundamentally has to do with a fair distribution of benefits from health and social development; while at the same time maintaining privacy and security, which makes it a hard proposition. Consideration of equity comes unto the picture when deciding the ethics. The global requirement at the moment is poverty reduction, energy, water, and health problems, and biodiversity maintenance. Although nanotechnology offers help in addressing these problems, the problem concerning access to the benefits of technology that takes into account the poorest and richest countries as well as people inside any country still remains. Equity involves both process and outcome. Equity, which is accepted as an expression of social justice, should be followed so as to have a fair distribution of benefits for health and social development not only for equality of access to healthcare, but also in accord with the needs of the individual in relation to the needs of all. Health equity should be such that equal resources are expended for each individual, i.e., supply equity; equal access to health services with equal quality of healthcare. Equity involves both process and outcome. Equity can be defined in different ways such as: (i) Equal status of health for all, which is eutopic, encompassing equal quality of healthcare and equal access to health services; (ii) Equal resources expended for each individual, i.e., supply equity; (iii) Care according to needs, i.e., demand equity; and also (iv) Equal resources expended for each case of a particular condition as well as equal healthy life gained per dollar expended.

(xi) *Cultural Diversity:* the question arises whether the ethical principles are universal or diverse and capable of being modified according to prevailing situations. Culture, which encompasses traditions, customs, and religious beliefs, has been deeply rooted in people for a very long time. Hence, the basic principles of ethics have to be relevant to all but with some degree of flexibility in applying them. Maybe globalization or accepting cultural diversity will succeed in bridging the gap between different cultures.

8.3.2 Privacy

Nanotechnology-based surveillance devices will surely be cause for concern; as hidden or invisible microphones or camera will affect privacy. There are an enormous number of innovations that provide wireless monitoring for healthcare so that providers and patients can exchange data when necessary. Therefore, a discussion on privacy and security is needed. Privacy concerns related to nanotechnology-based surveillance devices are indeed significant. With the increasing sophistication of surveillance technology, there is a growing need for robust legal and ethical frameworks to protect individuals' privacy rights. Several measures can be taken to address privacy concerns such as implementation of strict regulations and laws governing the use of nanotechnology-based surveillance devices. These regulations should outline permissible uses, data collection limits, and penalties for misuse. There should be transparency regarding the deployment and use of such technology. Authorities and organizations employing nanotechnology-based surveillance devices should be accountable for their actions, with mechanisms in place for oversight and review. Data minimization may assist in this; for example, collection of data should be minimized to only what is necessary for the intended purpose. Additionally, data should be anonymized whenever possible to protect individuals' identities. To safeguard the collected data from unauthorized access and breaches; robust encryption and security measures should be implemented. Moreover, the public should be educated and made aware of the capabilities and implications of nanotechnology-based surveillance devices. This can help individuals make informed decisions about their privacy and understand their rights. Establishing ethical guidelines for the development and use of these surveillance devices can ensure that they are used responsibly and ethically. Finally, given the global nature of surveillance technology, international collaboration is crucial to develop common standards and best practices for protecting privacy in the age of nanotechnology-based surveillance. However, despite these measures, ensuring complete privacy protection in the face of advancing surveillance technology remains challenging. Constant vigilance and adaptation of policies and technologies will be necessary to address emerging privacy threats effectively.

8.3.3 Security Concerns

There are already nanotechnology-based weapons like mini-nukes and mosquito-like robot weapons, invisible nano-needle bullets, new nano spies, bullet cameras (TNO concept), landmines, and improvised explosive devices [9]. These developments have raised concerns about whether these will ensure new ways of security or will be a potential "arsenal of techno-terrorism or even nano-terrorism" [10]. Regulating them leads to the complex questions of who will regulate them, how to regulate them, and how much transparency will be required in government and private nanotechnology initiatives to elude or evade misuses.

The scientific community has to open up to having a public debate about the social choices involved in developing a specific technology. This accessibility will enable us to better understand the multiplicity of positions on nanotechnologies and the reasons why these positions seem inaccessible to any form of argumentation.

8.3.4 Environmental Concerns

Environmental concerns have already been discussed in this chapter regarding all the steps involved in addressing ethical issues. Nanotechnology, while holding great promise for various fields such as medicine, energy, and materials science, also presents potential risks to the environment. If released into the environment, either intentionally or unintentionally, they are bound to interact with ecosystems in unforeseen ways. For example, nanoparticles could accumulate in soil or water bodies, potentially affect ing organisms and ecosystems. Nanoparticles may enter the food chain and bioaccumulate in organisms, leading to potential biomagnification as they move up the food chain. This can result in higher concentrations of nanoparticles in organisms at higher trophic levels, with potential ecological consequences.

Some nanoparticles have been found to exhibit toxicity in living organisms. Their small size and large surface area relative to volume can enhance their reactivity and ability to interact with biological systems. This toxicity could impact various organisms, from bacteria and aquatic organisms to terrestrial plants and animals. If nanoparticles enter ecosystems and disrupt natural processes or key ecological functions, they could potentially affect ecosystem services essential for human well-being, such as water purification, pollination, and nutrient cycling. Because nanotechnology is still relatively new, the long-term environmental effects of nanoparticles are not yet fully understood. There may be unforeseen consequences that emerge over time as nanomaterials become more prevalent in various products and applications.

Some nanoparticles may persist in the environment for extended periods due to their stability and resistance to degradation. Additionally, their small size and unique properties may allow them to move through soil, water, and air more readily than larger particles, potentially spreading contamination over large distances.

Protecting the environment from potential risks associated with nanotechnology-based products is an important consideration in the responsible development and use of nanotechnology. Several measures are being implemented to help safeguard the environment; such as many countries having established regulatory frameworks to assess and manage the environmental risks of nanotechnology-based products. These regulations may include requirements for environmental risk assessments, labeling, and restrictions on the use of certain nanomaterials. Prior to the commercialization of nanotechnology-based products, comprehensive risk assessments are being conducted to evaluate their potential environmental impacts, keeping in mind ethical concerns. This involves studying the behavior, fate, and toxicity of nanomaterials in various environmental compartments, such as air, water, soil, and organisms. There are efforts underway to develop green nanotechnology, life-cycle assessment, sustainable design, product stewardship and education for public awareness. While these measures help mitigate environmental risks associated with nanotechnology, ongoing research and collaboration are essential to address emerging challenges and ensure the sustainable development and use of nanotechnology-based products. By integrating environmental considerations into the design, production, and regulation of nanotechnology, we can help protect the environment while harnessing the benefits of this transformative technology.

8.4 Governance and Policies Based on Ethics

It was through a time-consuming multi-stakeholder dialogue that a responsible nano code emerged [11]. The principles this code was built on were based on accountability and stakeholder engagement, environmental health and safety, wider societal and ethical impacts, and transparency and disclosure, creating the basis for a framework for responsibility in nanotechnology innovation that will complement other initiatives.

Governance and policies based on ethics are critical for ensuring fairness, accountability, and the protection of rights within societies. However, several important concerns associated with implementing such governance and policies are raised. Let us explore some of these.

8.4.1 Subjectivity

Ethics can be subjective, varying between cultures, religions, and individuals. What one group considers ethical, another might not. This subjectivity can make it challenging to develop universally acceptable governance and policies. Different cultures have distinct ethical norms and values. Implementing a single set of ethical policies across diverse cultural contexts can be difficult and may lead to cultural imperialism or the marginalization of certain groups. Therefore, policies should respect cultural norms and values. Ethical governance promotes inclusivity and celebrates diversity. Policies should promote social justice, bridging gaps related to income, education, and access to resources. Ethical governance emphasizes fairness and equal treatment for all citizens.

8.4.2 Complexity

Ethical considerations often involve complex moral dilemmas where different ethical principles may conflict with each other. Balancing competing interests and values while creating policies can be difficult. To solve the conflict of interest, policymakers must avoid situations where personal interests conflict with their duty to serve the public. Addressing conflicts of interest should not be avoided as it ensures fair and unbiased policy formulation.

8.4.3 Enforcement of Policy

Even if ethical principles are clearly defined in governance and policies, enforcing them can be challenging. Without effective enforcement mechanisms, policies may not be followed, leading to ethical violations. Often power dynamics comes into play. Those in power may manipulate ethical frameworks to serve their interests, leading to unethical behavior being justified or overlooked. This can result in inequality and injustice within societies.

8.4.4 Evolution of Ethics and Ethical Dilemmas

Ethics are not static; they evolve over time as societal norms change. Policies based on outdated ethical principles may become irrelevant or even harmful as societies progress. Governance often involves making difficult decisions where there is no clear ethical solution. Policymakers may face ethical dilemmas where they must prioritize certain values over others, leading to controversy and criticism. If transparency and accountability are kept in consideration by those involved in ethical governance it will ensure that decisions are made in the public interest and not for personal gain. Lack of transparency can lead to suspicion, erode public trust, cause distrust in government institutions, and undermine the effectiveness of ethical policies. Ethical governance requires transparency in decision-making processes.

Addressing these concerns requires ongoing dialogue, engagement with stakeholders, and a commitment to upholding ethical principles in governance and policymaking processes. It also necessitates flexibility to adapt policies as ethical standards evolve and the recognition of the inherent complexity of ethical decision-making. Therefore:

- Policies should be accountable to citizens, ensuring that those in power are answerable for their actions.
- Policymakers must avoid situations where personal interests conflict with their duty to serve the public.
- Policies should promote social justice, bridging gaps related to income, education, and access to resources.
- Policies must respect fundamental human rights, including freedom, privacy, and dignity; so as to ensure protection against discrimination and abuse.
- Ethical policies must consider the long-term impact on the environment; and balance the economic growth with ecological preservation, which is very crucial.
- Whistleblowers should be encouraged to report unethical practices to ensure accountability; and policies should protect whistleblowers from retaliation.
- Ethical corporate policies should prevent fraud, corruption, and exploitation, ensuring that responsible business practices benefit society as a whole.
- Policies must strike a balance between individual freedoms and collective well-being. Ethical governance should avoid extreme measures that harm either side.
- Policies should respect cultural norms and values. Ethical governance has to promote inclusivity and celebrate diversity.
- Ethical policies should prioritize sustainable development over immediate profits; balancing short-term goals with long-term consequences is essential.

In summary, ethical governance ensures that policies align with moral principles, benefit society, and uphold the common good.

8.5 Problems and Challenges Faced in Addressing Ethical Concerns

As far as emerging risk to health due to engineered nanoparticles are concerned, the risk due to inhaled nanoparticles tops the list; moreover, it also includes the ability of nanoparticles to penetrate normally inaccessible places [12]. Various research studies conducted over the last three decades have proved that inhaled particles <100 nm in diameter elicit a significantly greater inflammatory response in the lungs and are more harmful than an equivalent mass of larger particles. Their possible adverse effect on the environment is equally a challenge that is being faced due to the increased use of nanoparticles that are being developed and commercialized [14–16].

These findings have led to awareness of a nanomaterial's potential impacts on human health at every stage of its life cycle—from production to transport for use, to disposal, and even to recycling [17–19]. As per the constitution of the World Health Organization, the definition of health is "a state of complete physical, mental, and social well-being and not merely the absence of disease or infirmity." The approach to the challenge of combating health-related issues, is to use a set of principles to identify and research engineered nanomaterials so as to ensure their safe and responsible development and use. Maynard *et al.* [20] have suggested the use of three possible principles encapsulating emergent risk, plausibility, and impact. These principles will be the tools that allow for addressing potential risks from engineered nanomaterials at all stages. Closer scrutiny from a risk perspective will be required to provide a framework within which all types of nanomaterials and products containing them will be encompassed. However, implementing ethical concerns can be challenging due to the factors mentioned below.

8.5.1 Complexity

Even definitions of ethics itself tend to be more complex. Ethical considerations can be intricate and multifaceted, involving various stakeholders, perspectives, and potential consequences. It can be challenging to navigate through these complexities and arrive at decisions that satisfy everyone involved. Nanotechnology, particularly in the context of nanoparticles, presents several ethical challenges during implementation. Workplace decision-making is one of the reasons for creating complexity, because of conflicting interests, which usually involves competing interests such as profit vs. employee well-being. This makes ethical decision-making complex.

8.5.2 Subjectivity

Subjectivity is the idea that perception comes from a subject's point of view, rather than from external facts. It's often opposed to objectivity, which is the idea that knowledge is independent of the subject. Ethics often involve subjective judgments about what is right or wrong, which can vary depending on cultural, social, and personal beliefs. This subjectivity can make it difficult to establish clear guidelines or standards for ethical behavior; leading to conflicting values; for example, different stakeholders may have conflicting values or priorities, making it challenging to find common ground or reach consensus on ethical issues. Balancing competing interests while ensuring fairness and integrity can be a significant challenge. Limited resources may also influence ethical choices.

8.5.3 Cost and Resource Constraints

Implementing ethical practices may require significant investments of time, resources, and effort. Organizations may face pressure to prioritize short-term financial goals over long-term ethical considerations, especially in competitive or resource-constrained environments. Time constraints require quick decision-making in the workplace, leaving little time for reflective reasoning. Ethical choices made under time pressure may rely on intuition rather than thorough analysis. Allocating resources (such as budget and personnel) requires ethically balancing competing needs. Prioritizing ethical considerations while managing limited resources can be complex.

8.5.4 Changing Landscapes

Ethical considerations can evolve over time due to changes in technology, society, or cultural norms. Keeping up with these changes and adapting ethical frameworks accordingly requires ongoing vigilance and flexibility. Product developed in one place and technology transported to other parts of the globe is a challenge, because in a globalized world, organizations often operate across diverse cultural contexts, each with its own ethical norms and values. Navigating these cultural differences while maintaining consistent ethical standards across borders can be challenging. Implementing ethical concerns may require changes to established practices or processes, which can face resistance from stakeholders who are comfortable with the status quo. Overcoming this resistance and fostering a culture that values ethical behavior may require strong leadership and effective communication.

8.5.5 Lack of Awareness, Education and Training

In some cases, individuals or organizations may not fully understand the ethical implications of their actions due to a lack of awareness or education. Promoting ethical awareness and providing training on ethical decision-making can help address this challenge. Lack of proper ethics training can hinder employees' ability to navigate ethical dilemmas. It is essential to cultivate awareness and understanding of ethical principles. Even well-intentioned efforts to address ethical concerns can sometimes result in unintended consequences. It's essential to carefully consider the potential impacts of ethical decisions and take steps to mitigate any adverse effects. This demands that employees be encouraged to report unethical behavior, which can be challenging due to fear of retaliation. Creating a safe environment for whistleblowers is crucial concern.

8.5.6 Health and Environment-Related Challenges

Striking a balance between improving patient outcomes and minimizing potential harm or ethical wrongs can be difficult in healthcare quality improvement projects. It has to be ensured that ethical considerations do not overly burden healthcare staff while implementing changes. Human health and safety can be controlled by restricting exposure routes: Nanoparticles can enter the body through inhalation, ingestion, or skin contact; therefore, ensuring workplace safety and consumer protection is essential. Workers handling nanoparticles face unique risks due to exposure during production and research. Privacy concerns arise when nanoparticle-based sensors are used for surveillance, tracking, or monitoring individuals. Since the behavior of nanoparticles at the atomic and molecular level is not fully understood, uncertainty arises in risk assessment due to unknown effects. Predicting the long-term effects of nanoparticles on human health and the environment remains challenging. Moreover, assessing their potential toxicity is complex due to variations in size, shape, surface chemistry, exposure routes and response by different subjects or individuals. With the realization that health and the environment are interrelated, the ethics relevant to both of them has become the focus of attention in the last few decades, and has evolved through several phases such as traditional ethics, bioethics, health policy ethics, and ethics in relation to environment and human rights. Since nanoparticles can accumulate in ecosystems, affecting aquatic organisms and soil health, their persistence and potential bioaccumulation raise concerns about long-term environmental consequences.

8.5.7 Organizational Ethics

Even with established ethical guidelines, enforcing them consistently across an organization can be challenging. Shaping an ethical organizational culture requires ongoing effort and commitment from leadership; otherwise, there will be lack of enforcement; resulting in some common problems and challenges faced when integrating ethics into different areas. Addressing these challenges requires a proactive approach that involves ongoing dialogue, collaboration, and a commitment to ethical principles at all levels of an organization.

8.5.8 Legal and Regulatory Compliance

Ethical concerns often intersect with legal and regulatory requirements, but compliance with laws and regulations does not always guarantee ethical behavior. Navigating the complex relationship between legal obligations and ethical responsibilities can be challenging for organizations. For legal and regulatory compliance, the following factors regarding regulation standards have to be taken into account: (i) *Lack of Clear Guidelines*: This is because regulatory agencies struggle to keep pace with rapidly evolving nanotechnology; (ii) *Standardization:* Consistent safety standards for nanoparticles should be established, which is rather challenging; (iii) *Equity and Access*: Distribution of benefits should ensure equitable access to nanotechnology benefits across different populations and countries, which is a challenge; (iv) *Health Disparity*: Nanomedicine advancements should not exacerbate existing health disparities. Ethical considerations in nanomedicine clinical trials include ensuring informed consent, transparency, and understanding of unknown risks; and most tricky are (v) *Dual-Use Dilemmas*: This is especially true where military and civilian applications are concerned, since nanoparticles can have both beneficial and harmful applications (e.g., drug delivery vs. weaponization). Hence, striking a balance between promoting innovation and preventing misuse is complex. Equity with a fair distribution of benefits from health and social development should be achieved through the solidarity approach and the individual rights approach.

Certainly, implementing ethical concerns can be challenging in the various contexts mentioned above. Integrating ethics into various contexts involves navigating conflicting priorities, resource constraints, and organizational culture. Addressing these challenges requires ongoing commitment and a proactive approach to ethical decision-making. In summary, addressing ethical concerns related to nanoparticles requires interdisciplinary collaboration, robust risk assessment, and proactive policy development.

8.6 Concluding Remarks

With great speed and fanfare, the application of nanotechnology has sprouted up in many branches of science and technology. However, along with the innovations brought by nanodevices; ethical, legal and social implications have also raised their heads.

In this chapter, concerns pertaining to ethics governance and policies from workplace to consumer market have been touched upon. Privacy, security, health and the environment are taken into consideration as the basis for governance and policymaking based on ethics. The problems faced, i.e., medical, environmental, social, cultural, etc., when implementing these new developments, in addition to their irreversible and long-term impact, has forced the creation of credible scenarios that can frame future decision-making processes. The ethical issues relating to nanotechnology are extremely critical and difficult to approach because it is hard to predict its development due to the lack of knowledge about its use in various situations. Therefore, the promise of nanotechnology can only be fulfilled in the future after observing its behavior in practical situations.

References

1. Oberdörster, G. and Yu, C.P., The carcinogenic potential of inhaled diesel exhaust: A particle effect? *J. Aerosol Sci.*, 211, S397–S401, 1990.
2. von Schomberg, R., Prospects for technology assessment in a framework of responsible research and innovation, in: *Technikfolgen abschätzen lehren: Bildungspotenziale transdisziplinärer Methoden*, M. Dusseldorp, and R. Beecroft (Eds.), Vs Verlag, Wiesbaden, 2011.
3. EC, Commission Recommendation of 07/02/2008 on a code of conduct for responsible nanosciences and nanotechnologies research. Brussels, Belgium, Commission of the European Communities, 2008.
4. Bensaude-Vincent, B., Opening the Field of Nanoethics: Editorial Introduction to the Special Issue on Bionanoethics, II. HYLE. *Int. J. Philos. Chem.*, 16, 1, 1–2, 2010.
5. Madhuri, S., Are we ignoring the considerations for the dark-side of nanotechnology. *J. Nanomed. Res.*, 3, 2, 1–2, 2016. NIA, 2008. The Responsible Nano Code – Update May 2008. Nanotechnology Industry Association. <http://www.nanotechia.org/activities/responsible-nano-code>.
6. Chetna, S. and Madhuri, S., Nanoparticles & the dark-side of nanotechnology concerning health-care. *Der Pharm. Lett.*, 6, 4, 365–375, 2014.

7. Reeser, D., Natural versus synthetic chemicals is a gray matter. *Guardian*, 10 April 2013, http://blogs.scientificamerican.com/guest-blog/2013/04/10/natural-vs-synthetic-chemicals-is- a-graymatter/.
8. Boseley, S., Concerns over chemicals in food packaging misplaced, say scientists. *Guardian*, 19 February 2014, http://www.theguardian.com/lifeandstyle/2014/feb/19/chemicals- food-packagingscientists.
9. Sharon, M., Rodriguez, A.S.L., Sharon, C., Gallardo, P.S., *Nanotechnology in the Defense Industry: Advances, Innovation, and Practical Applications*, Scrivener Publishing, Wiley, USA, 2019.
10. Mnyusiwalla, A., Daar, A.S., Singer, P.A., 'Mind the gap': science and ethics in nanotechnology. *Nanotechnology*, 14, R9–R13, 2003.
11. NIA, The Responsible Nano Code – Update May 2008. Nanotechnology Industry Association. <http://www.nanotechia.org/activities/responsible-nano-code> (accessed 12.10.15), 2008.
12. Ferin, J., Oberdörster, G., Penney, D.P., Soderholm, S.C., Gelein, R., H.C., P., Increased pulmonary toxicity of ultrafine particles. 1. Particle clearance, translocation, morphology. *J. Aerosol Sci.*, 21, 3, 381–384, 1990.
13. Oberdörster, G., Toxicology of ultrafine particles: *in vivo* studies. *Philos. Trans. R. Soc. Lond. A*, 358, 1775, 2719–2740, 2000.
14. Maynard, A.D., Nanotechnology: the next big thing, or much ado about nothing? *Ann. Occup. Hyg.*, 51, 1–12, 2007.
15. Schulte, P.A., Geraci, C.L., Murashov, V., Kuempel, E.D., Zumwalde, R.D., Castranova, V., *et al.*, Occupational safety and health criteria for responsible development of nanotechnology. *J. Nanopart. Res.*, 16, 1, 2014.
16. Sharma, V.K., Ma, X., Lichtfouse, E., *et al.*, Nanoplastics are potentially more dangerous than microplastics. *Environ. Chem. Lett.*, 21, 1933–1936, 2023. https://doi.org/10.1007/s10311-022-01539-1.
17. Klöpffer, W., Curran, M.A., Frankl, P., Heijungs, R., Köhler, A., Olsen, S.I., Nanotechnology and Life Cycle Assessment. *Synthesis of Results Obtained at a Workshop*, Project on Emerging Nanotechnologies, Washington, DC, 2007.
18. Gottschalk, F. and Nowack, B., The release of engineered nanomaterials to the environment. *J. Environ. Monit.*, 13, 5, 1145–1155, 2011.
19. Westerhoff, P. and Nowack, B., Searching for global descriptors of engineered nanomaterial fate and transport in the environment. *Acc. Chem. Res.*, 46, 3, 844–853, 2013.
20. Maynard, A.D., Chapter 1 - Challenges in Nanoparticle Risk Assessment, Editor(s): Gurumurthy Ramachandran, in: *Micro and Nano Technologies, Assessing Nanoparticle Risks to Human Health*, pp. 1–19, William Andrew Publishing, 2011, ISBN 9781437778632, https://doi.org/10.1016/B978-1-4377-7863-2.00001-7.

9

Bioethical Questions: Ethically Assessing Toxicology and Human Health

As our own species is in the process of proving, one cannot have superior science and inferior morals. The combination is unstable and self-destroying

Arthur C. Clarke

Abstract

This chapter explores bioethical concerns in nanotechnology. Bioethics addresses the moral and social implications of advances in medicine and healthcare, integrating law, science, and philosophy. International organizations like WHO, UNDRR, IRGC, and ISO, along with national bodies such as the EPA, FEMA, and NIOSH, guide ethical risk assessments. These organizations support frameworks to manage health risks, ensure equitable benefits, and address dual-use dilemmas. The chapter highlights the global collaboration required to implement ethical oversight and protect human health amid rapid nanotech advancement.

***Keywords*:** Bioethics, WHO, IRGC, UDRR, ISO, EPA, FEMA, NIOSH

9.1 Introduction

In this chapter, concerns and questions related to bioethics are discussed. Bioethics is the interdisciplinary study of moral and ethical questions related to human health and well-being. It explores the ethical implications of advances in health-related sciences such as biology, chemistry, and medicine. Bioethics brings together ideas from philosophers, physicians, lawyers, scientists, and public health professionals to address complex issues arising from medical research, clinical practice, and technological innovations. These questions often involve overlapping concerns from diverse fields, including life sciences, biotechnology, public health, medicine,

Madhuri Sharon. Risk, Safety, and Ethical Concerns in Nanotechnology, (187–222) © 2025 Scrivener Publishing LLC

public policy, law, philosophy, and theology. As technology continues to advance rapidly, bioethics plays an increasingly important role in shaping responsible practices and policies.

According to Beauchamp and Childress [1], there are four commonly accepted principles of healthcare ethics, they are: (i) Principle of respect for autonomy, (ii) Principle of nonmaleficence, (iii) Principle of beneficence, and (iv) Principle of justice.

New nanomedicines, biomedical procedures based on nanotechnology, and ways of altering plants and animals are bringing benefits to millions of people. However, these same innovations also have the potential to bring harm or to raise other kinds of ethical questions about their appropriate use. Bioethical questions related to nanotechnology and toxicology often revolve around the potential health risks and benefits associated with the use of nanomaterials in various applications. The intersection of nanotechnology, toxicology, and human health raises several bioethical questions. The questions discussed below, highlight the complex intersection of science, technology, ethics, and society in the context of nanotechnology and toxicology, emphasizing the need for multidisciplinary approaches to address emerging challenges responsibly.

9.2 Organizations Involved in Ethically Assessing Health Risks

There are several organizations and institutions globally focus on risk assessment and uncertainty analysis across various fields. Some of the notable ones that are directly or indirectly related to nanotechnology can be categorized as international organizations, government agencies of a country, financial and economic organizations, academic and research institutions, and industry and private sector.

9.2.1 International Organizations

There are various international bodies involved in promoting safe health (Figure 9.1).

Figure 9.1 International organizations for ethical assessment of health risks used by government agencies.

9.2.1.1 World Health Organization (WHO)

As its name indicates, the World Health Organization (WHO) works worldwide to promote health, keep the world safe, serve the vulnerable, protect people from health emergencies, and provide a further billion people with better health and well-being. This means they are involved in all aspects that directly or indirectly affect human health. The WHO constitution states: "*Health is a state of complete physical, mental and social well-being and not merely the absence of disease or infirmity.*" In the Parma Declaration on Environment and Health (2010) conducted by WHO, it has been agreed that more research is needed to assess the health risks and benefits on the use of nanomaterials and on the improvement of methods. Risk assessments related to public health are conducted by WHO, including disease outbreaks and global health threats. There is a "WHO Guidelines to Protect Workers from Nanomaterials" [2].

9.2.1.2 United Nations Office for Disaster Risk Reduction (UNDRR)

The UNDRR is part of the UN Secretariat. It's the UN's focal point for coordinating efforts to reduce disaster risk. It was established in 1999, with a mission to implement the International Strategy for Disaster Risk Reduction (ISDR) and the Sendai Framework. It works to make societies and countries more resilient by supporting the development of policies and actions. While the UNDRR's work doesn't specifically address risks from nanotechnology, it does collaborate on a special report about technology for disaster risk reduction.

9.2.1.3 International Risk Governance Council (IRGC)

The IRGC is accredited to the United Nations Environment Programme (UNEP). The IRGC is an independent foundation based in Switzerland whose purpose is to help the understanding and management of important, emerging global risks. In 2005, the IRGC decided to address the risk governance of nanotechnology as an emerging technology that offers both potentially enormous benefits but also presents significant challenges to health, government, industry, and society at large. The IRGC focuses on the governance of risks and uncertainties, particularly those with global implications. The general IRGC model encompasses five elements: Pre-assessment; Appraisal; Characterization and Evaluation; Management; and Communication. Risk communication is essential for establishing the two-way dialogue needed at all stages of the risk handling process.

9.2.1.4 International Organization for Standardization (ISO)

Though the ISO is not part of the United Nations, the two organizations do have a partnership. The ISO develops and publishes international standards, including those related to risk management (e.g., ISO 31000).

Every country has their own specific government agencies to govern national activities. For example, the FDA is a member agency in the National Nanotechnology Initiative (NNI), a federal research and development program established to coordinate the multi-agency efforts in nanoscale science, engineering, and technology. Examples of government agencies in the USA are mentioned in (Figure 9.2).

9.2.1.5 Environmental Protection Agency (EPA) – USA

The US Environmental Protection Agency (EPA) conducts risk assessments related to environmental health and safety, which assess the potential environmental impact of nanomaterials and their potential effects on human health. The EPA's research focuses on *occurrence* by characterizing the presence of nanomaterials; *transport* by examining how nanomaterials move and change in the environment; and *exposure* by estimating

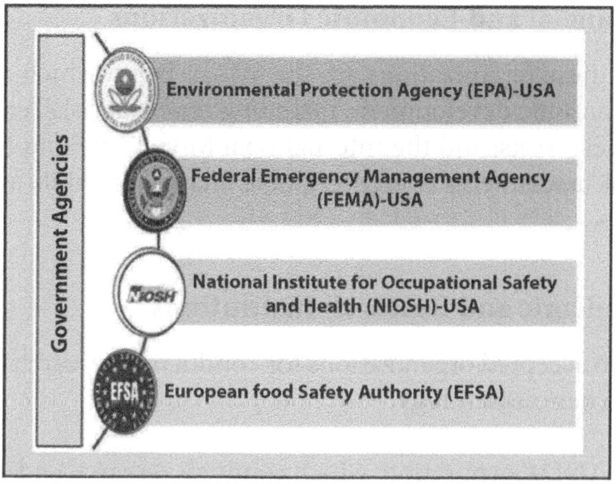

Figure 9.2 Government agencies involved in governing national activities.

how much humans may be exposed to nanomaterials. Moreover, the EPA develops science and processes to evaluate the human health and ecological safety of products containing nanomaterials to estimate their potential environmental impact. The EPA collaborates across federal and state governments, industry, and the international community on nano-related research.

9.2.1.6 Federal Emergency Management Agency (FEMA) – USA

The risk assessment of natural and man-made disasters are the focus of FEMA.

9.2.1.7 National Institute for Occupational Safety and Health (NIOSH) – USA

Risk assessments related to workplace safety and health are conducted by NIOSH.

9.2.1.8 European Food Safety Authority (EFSA)

Risk assessments related to food safety in the European Union are conducted by EFSA.

9.2.2 Financial and Economic Organizations

Included in these organizations are the World Bank, which assesses risks related to economic development, including financial risks, environmental risks, and social risks; and the International Monetary Fund (IMF), which conducts risk assessments related to global financial stability and economic uncertainties.

9.2.3 Academic and Research Institutions

There are two accepted organizations for conducting research and academics related to nanotechnology:

(i) RAND Corporation, which conducts research and analysis on a wide range of risk-related topics, including national security, health, and public policy.
(ii) Institute for Risk and Uncertainty, University of Liverpool, which focuses on advanced methodologies for understanding and managing risk and uncertainty in engineering and science.

9.2.4 Industry and Private Sector

These organizations utilize a range of methodologies and frameworks to assess and manage risks and uncertainties in their respective domains. They include (i) Deloitte, which provides risk assessment and management services for various industries. (ii) PricewaterhouseCoopers (PwC), which offers services related to risk assurance, including enterprise risk management and regulatory compliance; and (iii) Environmental Resources Management (ERM), who specialize in risk assessment related to environmental and sustainability issues.

9.3 How to Assess Risk and Uncertainty

Prior to answering this question, let us look at the question of whether current risk assessment methodologies for nanoparticles are adequate. What considerations should be taken into account in framing the risk assessment? Which factors of exposure for risk assessment methodologies need to be specified to address risks and hazards related to nanoparticles? And what should be done to improve the risk assessment of nanoparticles?

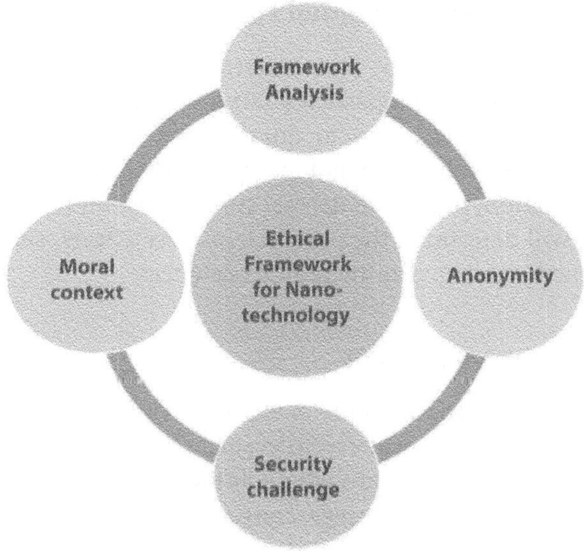

Figure 9.3 Framework for ethical analysis of risks in nanotechnology.

Risk assessment is a process that identifies and analyzes potential events that could negatively impact people, assets, or the environment. It also determines the likelihood and consequences of these events, and the tolerance for them. The results of a risk assessment can be expressed quantitatively or qualitatively. Assessing risk and uncertainty related to nanotechnology involves navigating complex challenges due to the unique properties and rapid development of nanomaterials, understanding potential hazards associated with nanomaterials, as well as uncertainties related to their behavior and impact on health and the environment (Figure 9.3).

9.3.1 Risk Assessment Frameworks

For risk assessment in nanotechnology, can we adopt the existing traditional models? The answer rests on the fact that the traditional risk assessment models may not fully apply to nanotechnology. There is a need to adapt existing frameworks to account for nanoparticle-specific factors. Experts from various fields (toxicology, epidemiology, engineering, etc.) have to assess risks comprehensively involving a multidisciplinary approach.

9.3.2 Identification of Potential Hazards

Thorough characterization of nanomaterials to determine the physico-chemical properties of nanoparticles (size, shape, composition, surface charge, surface chemistry, etc.) in question, as well as toxicity studies, should be conducted to identify potential hazards. The potential exposure pathways (inhalation, ingestion, dermal contact) for both humans and the environment will determine the potential adverse effects, such as toxicity, inflammogenicity, and environmental persistence.

9.3.3 Exposure Assessment

As previously mentioned, the route of exposure (inhalation, ingestion, dermal exposure) has to be considered to evaluate the likelihood and extent of exposure to the nanomaterial throughout its life cycle (manufacture, use, disposal); for this, occupational settings, consumer products, and environmental release pathways have to be recorded. This will help in considering and quantifying occupational and environmental exposure levels and durations.

9.3.4 Toxicity Assessment

The toxicity assessment determines the toxicological properties of the nanomaterial through *in-vitro* and *in-vivo* studies of factors such as dose, route of exposure, and duration of exposure. Toxicity levels assess the potential for adverse health effects, including acute and chronic toxicity, genotoxicity, and carcinogenicity. Toxicity depends on the dose-response relationships, which may vary with age of person, immunity level, health condition, etc. Knowing the threshold level for dose-response relationships helps in declaring the adverse effects. It requires extrapolation from high-dose animal studies to low-dose human exposure. At present, the most significant ethical issues relating to toxicity due to nanomedicine involve risk assessment, risk management, and risk communication.

9.3.5 Quantifying Uncertainty and Data Gaps

To address uncertainty and data gaps in hazard identification and exposure assessment, sensitivity analysis should be conducted to explore the impact of uncertain parameters such as variability in toxicity responses and exposure scenarios. Then, exposure and toxicity information should be integrated to estimate the potential risks associated with the nanomaterial and quantify risks using probabilistic approaches if possible.

9.3.6 Risk Management

Risk management (Figure 9.4) has to be done at five levels: (i) *Risk Characterization* using risk matrices to integrate hazard and exposure information. (ii) *Qualitative and Quantitative Risk Estimates* to provide both qualitative and quantitative risk estimates. (iii) *Continuous Monitoring and Research* to monitor emerging data and research findings to update risk assessments, management strategies; invest in ongoing research to fill knowledge gaps and reduce uncertainties regarding the health and environmental impacts of nanomaterials. (iv) *Precautionary Measures*, especially when data are limited, should be weighed for potential benefits against potential risks, which helps in balancing benefits and risks. (v) *Develop Strategies* to mitigate identified risks, such as engineering controls, personal protective equipment, or substitution with less hazardous materials, implementing exposure monitoring programs in occupational settings, and establishing regulations and guidelines for safe handling, use, and disposal of nanomaterials.

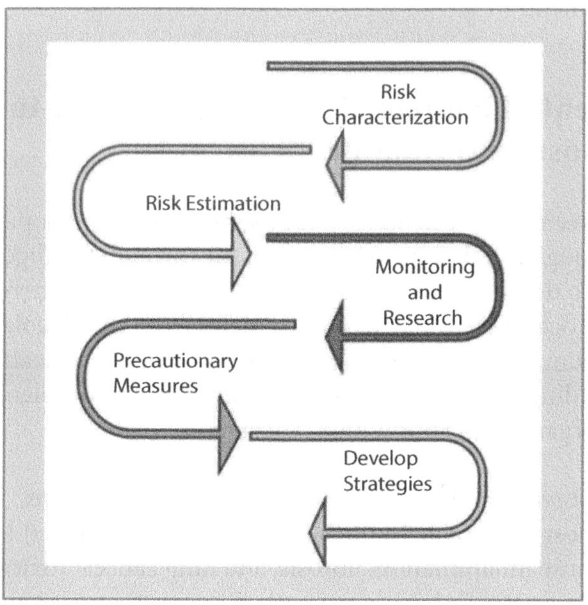

Figure 9.4 Levels of risk management process.

9.3.7 Ethical Considerations, Communication and Stakeholder Engagement

For ethical considerations, equity should be ensured by equitable distribution of risks and benefits. Transparency should be maintained by involving stakeholders in risk assessment and decision-making. For long-term implications, intergenerational equity and long-term consequences have to be considered. This will demand communicating risk assessment findings and uncertainties to relevant stakeholders, including workers, consumers, and regulators; and the transparency and collaboration among researchers, industry, and government agencies. Moreover, public concerns and perceptions related to nanotechnology risks through education and outreach efforts should be addressed. By following these steps, stakeholders can systematically assess and manage the risks and uncertainties associated with nanotechnology, ensuring the safe and responsible development and use of nanomaterials.

In summary, assessing risk and uncertainty in nanotechnology requires a multidisciplinary approach, transparency, and ethical considerations. Balancing innovation with safety is essential for responsible nanotechnology development.

9.4 Potential Human Health Impacts Due to Exposure to Nanoparticles

As we have seen in the previous chapters, exposure to nanoparticles (NPs) whether through inhalation, skin wounds, or through the digestive system, can percolate down to other parts of the human body. There is evidence for the translocation of inhaled nanoparticles from the alveolar spaces into the interstitium [3], which can have a variety of negative health effects and issues, including respiratory issues, cardiovascular problems, problems with other organs of the body, and cancer, etc.

(i) *Respiratory Issues:* Those with respiratory issues are the most vulnerable to further complications. Inhaled NPs can cause inflammation, fibrosis, and lung cancer. Toxicological observations have shown that exposure to inhaled NPs affects the early cellular/biochemical events and elicits an inflammatory response in the airways; which is measured by an increase in the number of neutrophils and in cytokine and chemokine levels resulting from oxidative stress pathways inside cells [4–6]. Moreover, inhaled NPs initiate an inflammatory response that may damage the epithelial cell layer as well as macrophages, leading to decreased phagocytosis [7]. Another possibility is the occurrence of neurogenic inflammation where airway nerve cells synthesize neurotransmitters that may affect white blood cells and epithelial tissue [8], causing inflammation that may reduce respiratory function and even cause increased susceptibility to infections.

(ii) *Cardiovascular Issues:* While studying the mechanisms of toxicity for nanoparticles, cardiovascular issues due to NPs have been noted. The adverse effects of NPs include heart and blood pressure and other cardiovascular effects [9–11]. Moreover, NPs have reportedly had effects on blood coagulation and systemic inflammation [12, 13].

(iii) *Neurological and Immune System Issues:* Chronic exposure to NPs can lead to neurotoxicity, which is associated with Alzheimer's, Parkinson's, and dementia. Nanoparticles, on the other hand, could have multiple health endpoints. They can interact with immune cells, potentially increasing susceptibility to autoimmune disorders, infectious diseases, and cancer. For example, inhaled nanoparticles might translocate to the central nervous system either through neural pathways in the olfactory system or through the body's circulatory systems (blood and lymphatic) after absorption in the lungs and consequent crossing of the blood-brain barrier. Campbell *et al.* [14] have noted that the inflammatory biomarkers, such as interleukin 1α (IL1α) and tumor necrosis factor α (TNFα), were increased in the brain of mice exposed to ambient air particulate matter compared to controls. Since one mechanism of toxicity of nanoparticles is likely to be induction of reactive oxygen species and the consequential oxidative stress in cells and organ, it warrants further studies of the induction of inflammatory cytokines. A relation with a variety of neurological diseases might be considered; and testing for interaction of nanoparticles with proteins and various cell types should also be considered as part of the toxicological evaluation.

(iv) *Reproductive Issues:* Since NPs can be toxic to fetuses and affect reproductive systems, there have been many standard tests done on the effect on the reproductive system, e.g., *Daphnia magna* reproduction test, earthworm (*Eisenia fetida* or *Eisenia andrei*) reproduction test for assessing the effects of chemicals in soil, and avian reproduction toxicity test. For reproductive effects, the Benchmark exposure limits for nanoparticles from the Institute for Occupational Safety and Health (referred to as the IFA) in Germany have suggested that for Nickel, cadmium-containing quantum dots, and chromium(VI) is $2 \times 10^7 p/m^3$ and for Beryllium, arsenic, zinc chromate it is $4 \times 10^7 p/m^3$. Whereas the British Standards Institute (BSI) in the United Kingdom suggested for Nickel, cadmium-containing quantum dots, chromium(VI), beryllium, arsenic, zinc chromate it is 0.1 × WEL (work exposure limits). McKee *et al.* [15] studied the impact of silver nanoparticles on the reproduction of Collembola, an invertebrate, and found it to be inhibited by a silver nanoparticle. Antimicrobial AgNPs are integrated into several products. During production, use, and disposal, they are emitted and enter into the environment *via* several pathways, soil is one of the sinks. Another NP that is being increasingly used in medicine and environmental remediation is iron oxide NP, which is a cause for concern as it has been found to affect the physiology and reproductive capacity of *Arabidopsis thaliana* at concentrations of 3 and 25 mg/L (Bobbin *et al.* 2015) [16]. According to Pérez-Hernández *et al.* [17], TiO_2 could induce cell damage related to exposure size and dose.

(v) *Cell and Organ Damage Issue:* Nanoparticles can cause irreparable cell damage through oxidative stress by creating free radicals. Certain nanoparticles can cause permanent cell damage, leading to organ injury and oxidative stress. Their small size and composition play a role in this toxicity. Moreover, NPs can damage almost any organ, including the lungs, liver, kidneys, nerves, and brain. There are reports of CNTs damaging cell membrane, leading to fatality [18]. Nowadays, metabolomic analysis, which is a powerful bioanalytical strategy to assess the toxicity of different NPs, is used to assess the toxicity induced by NPs *in vitro* [19]. Almost all parts of a cell have been found to be adversely affected by the NPs, which is bound to cause organ damage, e.g., cell membrane damage [20], mitochondrial damage [21, 22], cell cycle dysregulation [23],

DNA damage [24], genotoxicity [25], lipid peroxidation leading to changes in cell morphology [26], apoptosis [20] or necrosis [27], and metabolic changes [28].

(vi) *Cancer:* Use of NPs in theranostics, especially in the field of cancer is rapidly increasing (Puri 2010 and Sharon 2022) [29, 30]. NPs can cause DNA damage and trigger an inflammatory response, both of which are linked to carcinogenesis. According to Pope *et al.* (2002) [31] long term exposure to fine particles Lung cancer leading to cardio-pulmonary mortality is a major concern. *In vitro* test methods are being implemented as screening tools to assess potential toxicity mechanisms of NPs (Dechsakulthorn, *et al.*, 2008; Foldbjerg, *et al.*, 2011) [32, 33].

9.5 Mitigation of NP-Related Environmental Concerns Affecting Health

Prior to deciding the mitigation limit, the threshold level has to be decided. The level below which any toxicant does not show any observed adverse effects, considered as the limit or guideline or reference concentration, is established below this threshold. Exposures exceeding this threshold are considered to cause adverse health effects, and measures to mitigate or reduce exposures need to be taken. Mitigating nanoparticle-related environmental concerns requires a comprehensive approach to minimize risks and promote responsible use.

Environmental concerns related to nanoparticles arise from their widespread use and potential impact on ecosystems, including air, soil, water, plants, and aquatic life Nanoparticles being small in size allows them to penetrate cells and tissues, potentially causing harm to aquatic life, plants, and animals. They may accumulate in organisms over time. This bioaccumulation can disrupt ecosystems and affect food chains. Nanoparticles can enter water bodies through industrial processes, wastewater discharge, or runoff. Their presence in aquatic environments can alter water quality and harm aquatic organisms. Airborne nanoparticles can result from manufacturing, combustion, or other processes. Inhalation of these particles may impact human health and ecosystems. Nanoparticles can affect soil health, plant growth, and microbial communities. Their persistence in soil can alter nutrient cycles and impact soil organisms. The long-term environmental effects of nanoparticles are still not fully understood. Research is ongoing to assess their impact on ecosystems and biodiversity. Efforts are being made to balance the benefits of nanotechnology with responsible environmental stewardship.

Mitigating nanoparticle-related environmental concerns that affect health involves several approaches aimed at understanding, managing, and minimizing potential risks. Many strategies exist to achieve this. By implementing these strategies, stakeholders can work towards mitigating nanoparticle-related environmental concerns and minimizing their potential impacts on human health and ecosystems. The necessary strategies for mitigating the impact of nanoparticles are given below.

9.5.1 Monitoring and Assessment of Risk to Mitigate Effect of NPs

First and foremost, assessing the risks of nanoparticles (NPs) released into the environment and establishing regular regulatory monitoring accordingly, assists in assessing their impact on ecosystems, wildlife, and overall environmental health [34]. Assessment of risks helps to understand the potential impacts of particular nanoparticles on environmental and human health. The next step should be to establish regulations and guidelines specific to nanoparticles, considering their unique properties and potential risks. To establish regulations and to make appropriate policies; identifying and addressing potential risks associated with nanoparticle exposure is a must. It will prevent adverse health effects from inhalation, ingestion, or skin contact. Here, it is imperative to consider overall environmental health impacts.

9.5.2 Safe Nanoparticle Design and Practices to Mitigate Effect of NPs

The design and production of safer nanoparticles (NPs) with reduced toxicity and improved biocompatibility should be encouraged; and green chemistry principles should be implemented to minimize hazardous byproducts during nanoparticle synthesis. Such proactive measures can safeguard both human health and the environment. Also, effective methods for detecting and monitoring nanoparticles in the environment should be implemented, which will require investment in advanced detection technologies to trace and quantify nanoparticles in various environmental matrices. Moreover, further environmental fate and transport studies of the behavior of nanoparticles in different environmental media (air, water, soil) to predict their fate and transport pathways will help in understanding how nanoparticles interact with organisms and ecosystems over time. Biocompatibility, degradation rates, and environmental persistence during design have to be major considerations [35]. The National Institute for Occupational Safety and Health (NIOSH) has developed a detailed research strategy addressing the evaluation, characterization, and management of workplace health risks associated with engineered nanomaterials [36]. However, since nanoparticles pose a high degree of uncertainty, it can be difficult to perform a risk assessment to mitigate impacts through the redesign of materials, products, or technologies. Therefore, nanotechnologies and nanomanufacturing processes should be developed and disseminated in forms or ways that could help mitigate these impacts. If nanomaterials are assessed *in silico* before they undergo manufacturing, it will empower developers to validate the mitigation or eradication of properties of concern, and enhance the safety profile of nanomaterials.

9.5.3 Environmental Remediation and Nano-Bioremediation to Mitigate Effect of NPs

Once the potential adverse effects of nanoparticles (NPs) are thoroughly understood, the next step will be to explore remediation using nanoparticle-based technologies for environmental purposes, such as nanoparticle-enhanced water treatment or soil remediation. Mitigating nanoparticle-related environmental concerns requires a multifaceted approach such as (i) Bioremediation strategies to minimize the toxic effects of nanoparticles; in which techniques like phytoremediation can help reduce nanoparticle impact on ecosystems; and (ii) Nano-bioremediation by combining nanotechnology and bioremediation for efficient, eco-friendly remediation, with the aim of reducing contaminant concentrations to low risk-based levels. A combination of nanotechnology and bioremediation is needed to achieve more efficient, eco-friendly pollutant reduction. It will reduce contaminant concentrations while alleviating environmental impacts [37]. Predicting how nanoparticles will react under different environmental conditions has remained difficult.

The long-term effects of nanomaterials on soil microbiota and ecosystem functioning are not well understood and need to be addressed to fill in the research gaps. To mitigate possible risks associated with the use of nanomaterials in agriculture, interdisciplinary research collaborations and regulatory frameworks are needed [38]. In addition, risks of toxicity and other hazards of nanowaste are not well understood. Nano-QSAR models can predict the fate and behavior of nanomaterials in soil and water systems, helping to identify possible risks and develop strategies for mitigation [39]. Proper precautions can help to mitigate and manage the potential public health and environmental risks of nanomaterials without sacrificing their medical, social, and economic benefits. Finally, it can be said that Nano-QSAR (as mentioned in Chapter 6) aids in identifying promising nanomaterials while mitigating adverse effects on the environment and human health. That can be done by portraying the relationship between nanoparticle properties and toxicity [40].

9.5.4 Education and Public Awareness to Mitigate Effects of Nanoparticles

Increased public awareness and education among stakeholders (general public, industry, policymakers) about safe handling, disposal, and risk mitigation, as well as nanoparticle-related environmental and health risks, is a must. This will promote responsible use and disposal practices for nanoparticles. Another strategy at the base level is to carry on collaborative research and innovation by fostering interdisciplinary research collaborations to address gaps in knowledge regarding nanoparticle behavior, toxicity, and mitigation strategies. Innovation in green nanoparticle synthesis and sustainable nanotechnologies should be encouraged. Life cycle sustainability analyses should be conducted to evaluate the overall environmental impact of nanoparticles, from production to disposal. Sustainable practices that minimize nanoparticle waste and maximize recycling or safe disposal should be promoted, along with responsible practices in research, manufacturing, and product development. Mitigation demands require countries to work together on a global scale to establish harmonized standards for nanoparticle safety and environmental impact assessments. Collaboration on global research initiatives to address nanoparticle-related environmental concerns comprehensively is also required. Understanding and mitigating potential risks are fundamental to adopting responsible and sustainable practices.

9.5.5 Regulation and Policies to Mitigate Effects of Nanoparticles

Responsible use of nanoparticles benefits both technology and the environment. However, to prevent adverse health effects arising from their inhalation, ingestion, or skin contact, regulations and policies need to be established to identify and mitigate potential risks associated with nanoparticle exposure [41]. Reasonable precautions should be taken by policymakers by using the best available evidence and existing laws to regulate engineered nanomaterials and by supporting additional research to assess their risks.

9.6 Equitable Distribution of the Benefits of Nanotechnology

Equitably distributing the benefits of nanotechnology involves several key strategies to ensure that all segments of society can access and benefit from advancements in this field. This is a crucial consideration to ensure that this transformative field serves humanity as a whole.

It is imperative to do inclusive research and development by diverse research teams. Encouraging diversity in research teams will ensure a variety of perspectives and ideas in nanotechnology development. Promoting global collaboration is crucial because it allows sharing knowledge and resources, especially with developing countries, and offers open access support of research findings and data to allow scientists worldwide to contribute and benefit. By fostering international collaboration and knowledge exchange partnerships between developed and developing countries, research institutions, as well as industry players, will be encouraged. It will facilitate the transfer of nanotechnology expertise and solutions to regions that need them most, which can help bridge the gap between different parts of the world [42].

Hence, the technology used to develop affordable nanotechnology solutions should be accessible to low-income populations. To ensure that advancements in nanotechnology are not solely for the wealthy or developed nations, efforts should be made to share knowledge, resources and technologies across socioeconomic boundaries. For this, investments are needed in infrastructure to support the deployment and maintenance of nanotechnology, particularly in rural and underserved areas; as well as offers of subsidies or grants to startups and companies that focus on creating accessible and affordable nanotech products. So as to ensure that advancements in nanotechnology are not solely for the wealthy or developed nations, efforts should be made to share knowledge, resources, and technologies across socioeconomic boundaries. For this to happen, education and awareness should be taken into consideration by conducting public awareness campaigns to educate people about the benefits and risks of nanotechnology. Investing in STEM (science, technology, engineering, and mathematics) education is important, particularly in underrepresented communities, to build a knowledgeable workforce. And training programs should be provided for workers to adapt to new technologies and industries emerging from nanotechnology. Their perspectives and needs should guide research and application priorities to empower disadvantaged communities through capacity-building in nanotechnology research and application [43].

Focus is needed on healthcare and environmental applications that address global health issues, such as infectious diseases, cancer, and chronic conditions, ensuring these innovations reach underserved populations. Use of nanotechnology to develop solutions for environmental challenges, such as pollution control and renewable energy, benefit communities globally. Developing sustainable nanotechnology practices that do not harm the environment or human health also has to occur.

Regular monitoring and evaluation are requirement for impact assessment: Social, economic, and environmental impacts of nanotechnology are to be regularly monitored and assessed to ensure that benefits are distributed equitably. A continuous feedback loop for communities to voice concerns and suggestions regarding nanotechnology applications has to be set up to ensure that continuous improvement takes place by adapting and improving policies and practices based on ongoing monitoring and evaluation results.

Regulatory frameworks, like global safety standards to prevent misuse and ensure safe application of nanotechnology, are to be established along with ethical guidelines to address potential risks and societal impacts, ensuring technology is used responsibly. Regulatory frameworks have to be based on ethical considerations and develop frameworks for equitable benefit sharing. This involves ensuring that the benefits derived from nanotechnology research and commercialization are distributed fairly among stakeholders, including researchers, communities, and end-users.

Equitable patenting creates patenting policies that prevent monopolization and encourage sharing of technological advancements. To prevent monopolies and promote open access to knowledge and innovations, intellectual property rights has to be addressed.

Finally, inclusive policymaking and governance involve a wide range of stakeholders, including marginalized groups in policymaking processes related to nanotechnology. It is a must to implement policies to ensure fair distribution of nanotech benefits, such as equitable healthcare access and distribution of economic gains, forming international agreements to promote the equitable sharing of nanotechnology benefits across borders.

By adopting these strategies, the benefits of nanotechnology can be more equitably distributed, ensuring that advancements in this field contribute to the well-being of all people, regardless of their socioeconomic status or geographic location. In summary, harnessing nanotechnology for global good requires a concerted effort to ensure equitable access, affordability, collaboration, and ethical practices. Let's work together to maximize the positive impact of nanotechnology on society.

9.7 How to Ensure Human Exposure to Nanoparticles and Informed Consent

Ensuring that human exposure to nanoparticles is managed responsibly, especially in research and applications, involves several critical considerations. Any process should be conducted ethically and safely, which requires a comprehensive approach involving regulatory compliance, scientific rigor, and clear communication. Here are the key steps to ensure human exposure to nanoparticles and informed consent:

(i) *Risk Assessment and Mitigation:* This should be the first step, in which thorough preclinical studies to evaluate the safety of nanoparticles should be conducted. This includes *in-vitro* (cell culture) and *in-vivo* (animal) studies to understand toxicity, biodistribution, and potential side effects. Other factors to consider include toxicity, particle size, surface chemistry, and route of exposure (inhalation, ingestion, dermal contact). For risk management, strategies to mitigate identified risks should be developed. This may involve dose limitations, monitoring protocols, and emergency response plans. Risk mitigation can be achieved by implementing safety measures to minimize exposure. This includes using personal protective equipment (PPE), proper handling procedures, and engineering controls (e.g., fume hoods, containment systems).

(ii) *Continuous and Long-Term Monitoring of Ongoing Assessment:* This involves monitoring participants throughout the study for any adverse effects. It includes regular health checks and prompt reporting of any issues. Be prepared to adjust the study protocol in response to new safety data or participant needs. Health surveillance of individuals exposed to nanoparticles should be done over time. Early detection of adverse effects allows timely intervention. Environmental monitoring, i.e., assessing environmental exposure levels and potential impacts on ecosystems, should also be included.

(iii) *Informed Consent Process:* Clear communication has to be provided to potential participants with comprehensive information about the study, including its purpose, procedures, risks, benefits, and alternatives. Use language that is easy to understand. Ensure that the consent documentation form provides all necessary details and complies with legal and ethical standards. It should outline the nature and purpose of the study, specific procedures involving nanoparticle exposure, and potential risks and benefits. Participants should be allowed to freely discuss information and pose questions, and they should have the right to withdraw at any time without penalty. Ensure the participants fully understand the information before signing the consent form. Informed consent must cover research studies. In research involving human subjects, informed consent is essential. Participants must receive clear information about the study, including potential risks related to nanoparticles.

(iv) *Transparency:* Researchers should provide a detailed explanation of the study's purpose, procedures, benefits, and risks; and participants must voluntarily agree to participate based on this information. The study should maintain transparency in all aspects of the process, including any adverse effects reported. This helps in building public trust and contributes to the broader scientific understanding of nanoparticle safety. Data sharing should include published findings from scientific journals that have undergone a peer review to validate the study's methodology and outcomes. For this, researchers, workers, and healthcare professionals are to be trained and educated about nanoparticle safety. This includes understanding exposure routes, symptoms of exposure, and emergency response. Also, the general public should be actively informed about nanoparticles, their applications, and safety precautions. Encourage responsible use and informed choices. In addition to addressing the ethical concerns of a research study, community engagement to build awareness and understanding of the study is also important. This can help in addressing public concerns and ensuring community support.

(v) *Regulatory Compliance, Regulatory Frameworks and Ethical Considerations:* Guidelines are to be adhered to. Guidelines set by regulatory bodies, such as the FDA (Food and Drug Administration), EMA (European Medicines Agency), or equivalent national authorities, should be followed. These guidelines often specify safety testing, ethical considerations, and necessary approvals. Government regulations related to nanoparticle exposure should be established and enforced. These may include permissible exposure limits (PELs) and reporting requirements. Industries should adhere to established standards and guidelines for safe handling and exposure management. Ethics committee approval from an Institutional Review Board (IRB) or ethics committee should be obtained. These bodies review the study's ethical aspects, ensuring participant safety and rights. Benefit-risk balance should be weighed for the potential benefits of nanoparticle research or applications against the risks. Ethical decision-making should prioritize human well-being. Be especially cautious when working with vulnerable populations (e.g., children, pregnant individuals, the elderly). Their informed consent process requires additional sensitivity.

Remember that responsible management of nanoparticle exposure requires collaboration among researchers, policymakers, industry, and the public. By prioritizing safety and informed consent, we can harness the benefits of nanotechnology while minimizing risks. By following these steps, researchers can ensure that human exposure to nanoparticles is conducted ethically, safely, and transparently, with informed consent from all participants.

9.8 Regulatory Framework for Nanotechnology

The regulatory framework for nanotechnology should balance the promotion of innovation and the protection of public health, safety, and the environment. To ensure consistent regulation, the framework of nanomaterials and nanotechnology might include:

(i) Provide a clear and precise definition of nanomaterials and nanotechnology.
(ii) Determine the scope of regulations, including which industries and applications (e.g., medicine, cosmetics, food, electronics) are covered.
(iii) Develop risk assessment and management methodologies associated with nanomaterials, considering their unique properties, and implement measures to manage identified risks, including guidelines for safe handling, use, and disposal of nanomaterials.
(iv) Support research to understand the health, safety, and environmental impacts of nanomaterials. Create databases for collecting and sharing data on nanomaterial properties, uses, and safety.
(v) Ensure "interagency" in regulatory coordination by coordinating the work of various regulatory agencies (e.g., EPA, FDA, OSHA in the US) to avoid overlapping or conflicting regulations. Collaborate with international bodies to harmonize regulations and standards, facilitating global trade and cooperation.
(vi) Develop and adopt international standards for the testing, labeling, and certification of nanomaterials. Provide guidelines and best practices for the safe production, use, and disposal of nanomaterials.
(vii) Ensure transparency and public involvement in the regulatory process, making information about nanomaterial safety and regulatory decisions publicly available. Engage stakeholders, including the public, industry, and academia, in the regulatory process through consultations and advisory committees.
(viii) Ensure regulatory flexibility and adaptability by designing regulations that are flexible enough to accommodate rapid advancements in nanotechnology. Establish mechanisms for regularly reviewing and updating regulations based on new scientific evidence and technological developments.
(ix) Ensure compliance and enforcement of monitoring and inspection to be developed. Procedures for monitoring compliance with nanotechnology regulations and conducting

inspections has to be enforced. Implement enforcement mechanisms to ensure compliance, including penalties for non-compliance.

(x) Follow guidelines for ethical and social considerations to address ethical issues related to nanotechnology, such as privacy concerns, equity, and the potential for misuse. Also assess and mitigate the social impacts of nanotechnology, ensuring equitable access to its benefits.

(xi) Promote proper education and training programs to raise awareness about nanotechnology safety and regulatory requirements. Provide training for workers handling nanomaterials to ensure safe practices.

By incorporating these components, a regulatory framework for nanotechnology can effectively manage risks while fostering innovation and ensuring public trust.

9.9 How to Tackle the Dual-Use Dilemma of Nanoproducts

The dual-use dilemma associated with nanotechnology arises from its abundant potential for both beneficial and harmful applications. Let's look into nanotechnology's dual-use potential.

We know that nanotechnology operates at the nanometer scale, where familiar materials exhibit altered, unexpected, and amplified properties. These properties include tensile strength, viscosity, conductivity, antimicrobial effects, etc. The convergence of nanotechnology with information technology, biotechnology, and cognitive science opens up a vast range of possibilities in engineering, medicine, energy, and computing. While proponents celebrate nanotechnology's potential for human betterment, there are also significant risks associated with its development and dissemination [44]. Nanotechnology is now well-established in research and industry across more than 60 countries, and nanomaterials appear in everyday consumer products without much fanfare. Research breakthroughs and rapid development are outpacing our regulatory systems. The rate of innovation challenges our ability to keep up with safety assessments and risk management. Lessons from the life sciences can inform ethical and policy considerations related to nanotechnology's dual-use nature. We must consider both the positive applications (e.g., medical diagnostics, clean energy) and the potential misuse (e.g., surveillance, weaponry) of nanotechnology. Risk governance becomes crucial in managing dual-use scenarios, emphasizing transparency, informed consent, and responsible research practices. In summary, addressing the dual-use dilemma requires a balanced approach that maximizes the benefits of nanotechnology while minimizing its risks. Ethical frameworks, international collaboration, and ongoing monitoring are essential to navigate this complex landscape.

9.10 Ethical Boundaries of Privacy and Surveillance Regarding Nanotechnology

Guidelines are crucial when it comes to protecting individual rights and societal values; therefore, establishing ethical boundaries for privacy and surveillance concerns in relation to nanotechnology is important. Nanoethics explores the ethical implications of using nanotechnology for surveillance purposes, such as the development of nanoscale sensors and tracking devices. It emphasizes the need for robust privacy regulations and the protection of individual rights in the context of nanotechnology applications. Additionally, when considering nanotechnology's impact on privacy and surveillance, several other ethical boundaries and guidelines are relevant.

The key ethical considerations and boundaries include transparency and informed consent to maintain privacy and ensure that individuals are informed about the use of nanotechnology in surveillance. Transparency regarding data collection, monitoring, and tracking is essential. Open communication and public consultation are required for transparency. Therefore, maintaining transparency on how nanotechnology is used in surveillance and its implications for privacy; and engaging the public in discussions about these ethical implications and governance of its use in surveillance is crucial.

Informed consent should be obtained, which requires explicit consent from individuals before deploying nanoscale surveillance technologies. This ensures that individuals are fully informed about the use of nanotechnology in surveillance devices or systems and consent to their use. Participation in any program involving nanotechnology should be voluntary, with individuals having the option to opt out. People should understand how their data is being collected and used. Hence, data security and robust encryption and secure storage mechanisms are crucial. This protects data collected by nanosensors from unauthorized access, and collects only the data that is strictly necessary for the specified purpose. It also ensures that data collected is anonymized where possible to protect individual identities. As of today, protection measures implement robust security measures to protect data collected through nanotechnology from unauthorized access, breaches, or misuse. Clear policies are established for how long data can be retained and ensure it is securely deleted when no longer needed. Other protection measures include avoiding unnecessary long-term storage; limiting the retention of surveillance data to the necessary duration and purpose; clearly defining and limiting the purposes for which nanotechnology is used in surveillance to prevent misuse; and avoiding the use of nanotechnology for unnecessary or overly invasive surveillance that infringes on personal privacy.

Involve the public by letting them participate in discussions about nanoscale surveillance. Public input can shape policies and ethical guidelines. Open debates about the ethical implications of nanotechnology in surveillance are also to be encouraged.

For individual autonomy and privacy rights, consider the balance between surveillance for security purposes and individual privacy rights. Strive for a fair trade-off. Design nanoscale surveillance systems to minimize intrusiveness and avoid unnecessary invasion of personal space. Respect the autonomy of individuals by allowing them to make informed choices about their participation in surveillance activities involving nanotechnology. It is imperative that an individual has the right to opt out of surveillance systems without facing negative consequences. By adhering to ethical boundaries, the use of nanotechnology in surveillance can be guided in a way that respects privacy, protects individual rights, and promotes public trust. Ethical research and development should be conducted on nanotechnology in a manner that adheres to ethical standards and prioritizes privacy. Perform ethical impact assessments to evaluate the potential effects of new nanotechnology on privacy and surveillance before deployment. Ethical boundaries related to privacy and surveillance in nanotechnology require a delicate balance between security needs and individual rights. By addressing these considerations, we can harness nanotechnology's potential while safeguarding privacy and autonomy. Recognize that nanotechnology can have both positive and negative applications. Ethical decision-making should consider the dual-use potential. Hence, evaluate the risks associated with surveillance technologies and weigh them against societal benefits.

Establish regulatory bodies to oversee the deployment of nanoscale surveillance technologies. These bodies can set guidelines, monitor compliance, and address ethical concerns. Hold organizations and individuals accountable for any misuse of nanoscale surveillance systems. Create independent oversight bodies to monitor and review the use of nanotechnology in surveillance. Implement regular audits and assessments to ensure compliance with ethical standards and regulations. Ensure that nanotechnology is not used in ways that discriminate against individuals or groups based on race, gender, socioeconomic status, or other characteristics. Take steps to prevent biases in the design and deployment of surveillance systems using nanotechnology. Finally, adhere to laws and legal compliance. Ensure that the use of nanotechnology in surveillance complies with existing privacy laws and regulations. Advocate for the development of new legal frameworks to address emerging ethical and privacy concerns related to nanotechnology.

9.11 Balancing Long-Term Effects and Precautionary Principle Pertaining to NPs

For balancing long-term effects and precautionary principle pertaining to nanoparticles, the strategies to be followed should include (i) comprehensive risk assessment; (ii) understanding the precautionary principle and taking precautionary measures; (iii) research and monitoring so as to make science-based decision; (iv) innovation and alternatives; (v) competing priorities; (vi) excluding implausible dangers; (vii) transparency and communication; (viii) adaptive and precautionary regulations and standards; (ix) ethical considerations; and (x) stakeholder engagement. By implementing these strategies, it is possible to balance the potential long-term effects of nanoparticles with the precautionary principle, ensuring both innovation and safety are prioritized. In a concise way let's look at each of the strategy.

(i) *Comprehensive Risk Assessment:* This involves identification of potential hazards associated with nanoparticles, including toxicological and environmental risks. Evaluate the likelihood and extent of exposure to nanoparticles in various contexts (e.g., manufacturing, consumer use, disposal). Combine hazard and exposure information to characterize the overall risk.

(ii) *Understanding Precautionary Principle and Taking Precautionary Measures:* For this, apply the precautionary principle by taking preventive measures when scientific evidence about potential risks is uncertain but plausible.

Implement safety margins in the use of nanoparticles, erring on the side of caution when there is uncertainty about long-term effects. Support ongoing research to understand the long-term effects of nanoparticles on human health and the environment. Establish monitoring systems to track the presence and impact of nanoparticles in the environment over time. Conduct long-term epidemiological studies to detect potential health impacts of nanoparticles on populations. Establish stringent standards for the production, use, and disposal of nanoparticles to minimize potential risks. Consider precautionary bans or restrictions on the use of specific nanoparticles that pose significant unknown risks until more information is available. The precautionary principle is not an extreme or irrational approach. Instead, it provides guidance on how to address scientific uncertainty when dealing with emerging technologies. Official definitions adopted by the European Union and signatories of international environmental protection treaties emphasize its role in managing uncertainty [45].

(iii) *Research and Monitoring to Make Science-Based Decision:* The precautionary principle encourages decisions based on the current state of science. It requires procedures for scientific updates to stay informed. Rather than stifling innovation, it aims to ensure that new technologies are introduced with reasonable safety precautions.

(iv) *Innovation and Alternatives:* Encourage the development and use of safer alternatives to potentially hazardous nanoparticles. Promote the design of nanoparticles and nanomaterials with sustainability and safety in mind, reducing potential long-term risks.

(v) *Competing Priorities:* In cases with competing top priorities, the precautionary principle should be combined with other decision principles. Balancing long-term effects involves considering not only safety but also other societal needs.

(vi) *Excluding Implausible Dangers:* The principle should exclude potential dangers whose plausibility is too low to warrant meaningful precautionary action. Focusing on realistic risks ensures a balanced approach.

(vii) *Transparency and Clear Communication:* To maintain transparency in regulatory processes, communicate the rationale for precautionary measures and perform risk assessments clearly to the public. Share information about the long-term effects of nanoparticles and precautionary measures with international regulatory bodies to promote global safety standards.

(viii) *Adaptive and Precautionary Regulations and Standards:* Create adaptive regulatory frameworks that can evolve based on new scientific evidence and technological advancements. Regularly review and update regulations and guidelines to incorporate the latest research findings and risk assessments.

(ix) *Ethical Considerations:* Develop and adhere to ethical frameworks that prioritize public health, safety, and environmental protection in the development and application of nanoparticles. Ensure that the benefits and risks of nanoparticles are distributed equitably, avoiding disproportionate impacts on vulnerable populations.

(x) *Stakeholder Engagement:* Involve inclusive decision-making by engaging a broad range of stakeholders, including scientists, industry representatives, policymakers, and the public, in decision-making processes related to nanoparticle regulation. Increase public awareness and understanding of the potential risks and benefits of nanoparticles, encouraging informed participation in policy discussions.

In summary, achieving a balance between long-term effects and the precautionary principle of nanoparticles requires informed decision-making, ongoing scientific updates, and a pragmatic assessment of risks. By doing so, we can harness the potential of nanoparticles while safeguarding human health and the environment.

9.12 Fostering Global Collaboration in Research, Regulation and Governance of Nanotechnology

A multifaceted approach that leverages international cooperation, harmonization of standards, and inclusive stakeholder engagement is required in order to foster global collaboration in the research, regulation, and governance of nanotechnology, which is essential for harnessing the full potential of this transformative field while ensuring responsible and safe development. International research collaborations need international funding and grants that encourage cross-border research collaborations in nanotechnology in order to create multinational research initiatives and consortia that bring together scientists and institutions from different countries to work on common nanotechnology challenges. Moreover, promoting research exchange programs and international workshops to facilitate knowledge sharing and networking in order to establish international networks and partnerships for collaborative research projects across borders. This can create international networks of scientists, engineers, and policymakers to share knowledge, data, and best practices. Moreover, collaborative research grants can promote cross-disciplinary studies and address global challenges.

Another strategy is harmonization of standards and regulations. International bodies like the ISO (International Organization for Standardization) and IEC (International Electrotechnical Commission) are working to develop and harmonize standards for nanotechnology. Encourage regulatory agencies to collaborate on developing consistent regulatory frameworks that can be adopted globally, reducing disparities and facilitating international trade. Create and maintain global databases of nanomaterials and their safety profiles to inform regulatory decisions and ensure transparency. Work toward harmonizing regulatory frameworks related to nanotechnology. Allow consistent evaluation of common safety standards and guidelines for nanomaterials across different jurisdictions. Promote mutual recognition of safety assessments and approvals. This streamlines the approval process for nanotechnology products in multiple markets.

Information exchange and capacity-building can be achieved by organizing workshops, seminars and conferences. These platforms foster dialogue, knowledge exchange, and capacity-building, while simultaneously developing training programs for regulators, researchers, and industry professionals in order to enhance their understanding of nanotechnology safety, risk assessment, and ethical considerations. It will also create a platform for data sharing with transparency, which encourages open access to research findings, safety data, and risk assessments. Since transparency builds trust and facilitates collaboration, by creating global databases for nanomaterial properties, toxicity profiles, and exposure data, researchers and regulators can contribute and access shared information. Implementing international training programs and workshops to build capacity in nanotechnology research and regulation, especially in developing countries, fosters partnerships between universities and research institutions to develop joint educational programs and curricula in nanotechnology. Online platforms can be utilized for collaborative learning, sharing resources, and conducting virtual conferences and seminars.

Along with establishing international governance frameworks, it is important to provide a framework for the ethical and scientific issues involved with nanotechnology to address ethical, legal, and societal implications of nanotechnology, ensuring responsible development and use. The global public can be involved in discussions about the risks and benefits of nanotechnology through transparent communication and public consultations. Develop and promote international ethical guidelines for nanotechnology research and applications by addressing issues like privacy, security, and environmental impact. Involve the public in discussions about nanotechnology governance because public engagement ensures diverse perspectives and informed decision-making.

Facilitate collaborations between academia, industry, and governments to drive innovation and ensure that research findings are translated into practical applications. Form international consortia and alliances that include diverse stakeholders from different sectors to address common challenges and leverage collective expertise. Encourage multinational corporations and international organizations to invest in nanotechnology research and development. Establish international ethics committees to address ethical, legal, and social implications (ELSI) of nanotechnology, which can guide policy decisions. Develop global monitoring systems to track the development and impact of nanotechnology, ensuring that it is aligned with sustainable development goals. Establish standardized metrics and indicators for evaluating the progress and impact of nanotechnology initiatives on a global scale.

Conduct regular reviews and assessments of global nanotechnology collaborations to identify best practices, challenges, and areas for improvement. By implementing these strategies, stakeholders can create a collaborative and harmonized global environment that fosters innovation while ensuring the safe and responsible development of nanotechnology. Adopt an anticipatory governance approach. Identify potential risks early and develop strategies to address anticipated challenges. Collaborate on comprehensive risk assessments, considering environmental, health, and safety aspects. Share risk assessment methodologies and findings globally. Global collaboration in nanotechnology research, regulation, and governance requires concerted efforts, shared resources, and a commitment to responsible development. By working together, we can maximize the benefits of nanotechnology while minimizing risks for humanity and the environment.

9.13 Concluding Remarks

In this chapter, bioethical questions regarding toxicology and human health were addressed, which requires interdisciplinary collaboration, transparent communication, and a commitment to responsible development and use of nanotechnology. The potential risks of nanomaterial exposure pose a different concern because of the unique properties of nanoparticles that differ from their bulk counterparts. Human health is impacted due to duration of exposure to engineered nanoparticles. The potential short-term and long-term health impacts concerning inhalation, ingestion, or dermal contact are a bigger concern. We have to balance innovation with the precautionary principle when dealing with potential long-term health effects of nanoparticles, because we have an ethical obligation to protect future generations from unintended consequences. Long-term monitoring will guide the establishment of surveillance systems to monitor the long-term health effects of nanomaterial exposure in both occupational and environmental settings. The interactions between nanomaterials used in consumer products or industrial processes impact the biodiversity of ecosystems and human beings. Therefore, there is a need for ethical considerations in assessing and mitigating these impacts. Both researchers and manufacturers have ethical responsibilities in balancing technological advancements with environmental stewardship. Also, there is a demand for ensured informed consent when individuals are exposed to nanoparticles (e.g., during clinical trials or occupational settings); especially vulnerable populations (e.g., children, pregnant women), etc.

In this chapter, the dual-use dilemma (beneficial applications vs. harmful uses) has also been touched upon. For example, how can the potential dual-use nature of nanotechnology, where advancements intended for beneficial purposes may also have harmful applications, be ethically managed to minimize risks to society? A very clear ethical framework to guide decisions related to national security and nanotechnology has to be formulated. Ethical boundaries should be established when using nanoparticles for surveillance, tracking, or monitoring individuals.

Finally, at a governance and global level, there should be equitable access to nanomedicine and nanotechnology benefits across diverse populations among different socioeconomic groups to avoid exacerbating existing health disparities. Appropriate regulatory frameworks needed to ensure the safe development, manufacturing, and use of nanotechnology-based products while balancing innovation and public safety.

References

1. Beauchamp, T. and Childress, J., *Principles of Biomedical Ethics*, 7th Edition, Oxford University Press, New York, 2013.
2. John, H. and Murashov, V., WHO Guidelines to Protect Workers from Nanomaterials, December 15, 2017.
3. Oberdörster, G., Oberdörster, E., Oberdörster, J., Invited review: Nanotechnology: An emerging discipline evolving from studies of ultrafine particles. *Environ. Health Perspect.*, 113, 7, 823–839, 2005.
4. Salvi, S., Blomberg, A., Rudell, B., Kelly, F., Sandstrom, T., Holgate, S.T., Frew, A., Acute inflammatory responses in the airways and peripheral blood after short-term exposure to diesel exhaust in healthy human volunteers. *Am. J. Respir. Crit. Care Med.*, 159, 702–709, 1999.
5. Salvi, S., Nordenhall, C., Blomberg, A., Rudell, B., Pourazer, J., Kelly, F.J., Wilson, S., Sandstrom, T., Holgate, S.T., Frew, A., Acute exposure to diesel exhaust increases IL-8 and GRO-a production in healthy human airways. *Am. J. Respir. Crit. Care Med.*, 161, 550–557, 2000.
6. Ghio, A.J., Kim, C., Devlin, R.B., Concentrated ambient air particles induce mild pulmonary inflammation in healthy human volunteers. *Am. J. Respir. Crit. Care Med.*, 162, 981–988, 2000.
7. Renwick, L.C., Donaldson, K., Clouter, A., Impairment of alveolar macrophage phagocytosis by ultrafine particles. *Toxicol. Appl. Pharmacol.*, 172, 119–127, 2001.
8. Barnes, P.J., Neurogenic inflammation in the airways. *Respir. Physiol.*, 125, 145–154, 2001.
9. Peters, A., Wichmann, H.E., Tuch, T., Heinrich, J., Heyder, J., Respiratory effects are associated with the number of ultra-fine particles. *Am. Respir. Crit. CareMed.*, 155, 1376–1383, 1997b.
10. Wichmann, H.-E., Spix, C., Tuch, T., Wolke, G., Peters, A., Heinrich, J., Kreyling, W.G., Heyder, G., Daily mortality and fine and ultrafine particles in Erfurt, Germany. Part I: Role of particle number and particle mass. Research Report 98, Health Effects Institute, Cambridge, MA, 2000.
11. Penttinen, P., Timonen, K.L., Tiittanen, P., Mirme, A., Ruuskanen, J., Pekkanen, J., Ultrafine particles in urban air and respiratory health among adult asthmatics. *Eur. Resp. J.*, 17, 428–435, 2001.
12. Brown, J.S., Zeman, K.L., Bennett, W.D., Ultrafine particle deposition and clearance in the healthy and obstructed lung. *Am. J. Respir. Crit. Care Med.*, 166, 1240–1247, 2002.
13. Chalupa, D.C., Morrow, P.E., Oberdörster, G., Utell, M.J., Frampton, M.W., Ultrafine particle deposition in subjects with asthma. *Environ. Health Perspect.*, 112, 879–882, 2004.
14. Campbell, A., Oldham, M., Beceria, A., Bondy, S.C., Meacher, D., Sioutas, C., Misra, C., Mendez, L.B., Kleinman, M., Particulate matter in polluted air

may increase biomarkers of inflammation in mouse brain. *NeuroToxicology*, 26, 133–140, 2005.
15. McKee, M.S., Engelke, M., Zhang, X., *et al.*, Collembola reproduction decreases with aging of silver nanoparticles in a sewage sludge-treated soil. *Front. Environ. Sci.*, 5, 19, 2017, doi: 10.3389/fenvs.2017.00019.
16. Bombin, S., LeFebvre, M., Sherwood, J., Xu, Y., Bao, Y., Ramonell, K.M., Developmental and reproductive effects of iron oxide nanoparticles in Arabidopsis thaliana. *Int. J. Mol. Sci.*, 16, 10, 24174–24193, 2015.
17. Pérez-Hernández, H., Medina-Pérez, G., Valle-García, J.D., Vera-Reyes, I., Fernandez-Luqueno, F., Food and Agricultural Nanotechnology Chapter 2 Edible Crop Production by Nanotechnology: Is It a Sustainable Technology for Healthy Soil?, in: *Intellectual Property Issues in Nanotechnology*, C. Keswani (Ed.), CRC Press, Taylor & Francis Group, Boca Raton, London New York, 2020.
18. Sharon Madhuri, S., Shah, S., Maheshwar Sharon, T., Soga & R. Afre. Photocatalytic degradation of E. coli & S. aureus by multi-walled carbon nanotubes. *Carbon Lett.*, 8, 3, 184 - 190, 2007.
19. Awashra, M. and Młynarz, P., The toxicity of nanoparticles and their interaction with cells: an *in vitro* metabolomic perspective. *Nanoscale Adv.*, 5, 2674–2723, 2023, DOI: 10.1039/D2NA00534D (Review Article).
20. Zhang, L., Wu, L., Mi, Y., Si, Y., Silver nanoparticles induced cell apoptosis, membrane damage of Azotobacter vinelandii and Nitrosomonas europaea *via* generation of reactive oxygen species. *Bull. Environ. Contam. Toxicol.*, 103, 181–186, 2019.
21. Fu, P.P., Xia, Q., Hwang, H.-M., Ray, P.C., Yu, H., Mechanisms of nanotoxicity: Generation of reactive oxygen species. *J. Food Drug Anal.*, 22, 64–75, 2014.
22. Zhao, H., Chen, L., Zhong, G., Huang, Y., Zhang, X., Chu, C., Chen, L., Wang, M., Titanium dioxide nanoparticles induce mitochondrial dynamic imbalance and damage in HT22 cells. *J. Nanomater.*, 2019, 1–16, 2019.
23. Ramalingam, V., Revathidevi, S., Shanmuganayagam, T., Muthulakshmi, L., Rajaram, R., Biogenic gold nanoparticles induce cell cycle arrest through oxidative stress and sensitize mitochondrial membranes in A549 lung cancer cells. *RSC Adv.*, 6, 20598–20608, 2016.
24. Duan, J., Yu, Y., Li, Y., Yu, Y., Li, Y., Zhou, X., Huang, P., Sun, Z., Toxic effect of silica nanoparticles on endothelial cells through DNA damage response *via* Chk1-dependent G2/M checkpoint. *PLoS One*, 8, e62087, 2013.
25. Senapati, V.A., Kumar, A., Gupta, G.S., Pandey, A.K., Dhawan, A., ZnO nanoparticles induced inflammatory response and genotoxicity in human blood cells: a mechanistic approach. *Food Chem. Toxicol.*, 85, 61–70, 2015.
26. Shvedova, A., Castranova, V., Kisin, E., Schwegler-Berry, D., Murray, A., Gandelsman, V., Maynard, A., Baron, P., Exposure to carbon nanotube material: assessment of nanotube cytotoxicity using human keratinocyte cells. *J. Toxicol. Environ. Health Part A*, 66, 1909–1926, 2003.

27. Yuan, X., Nie, W., He, Z., Yang, J., Shao, B., Ma, X., Zhang, X., Bi, Z., Sun, L., Liang, X., Tie, Y., Liu, Y., Mo, F., Xie, D., Wei, Y., Wei, X., Carbon black nanoparticles induce cell necrosis through lysosomal membrane permeabilization and cause subsequent inflammatory response. *Theranostics*, 10, 4589–4605, 2020.
28. Shin, T.H., Seo, C., Lee, D.Y., Ji, M., Manavalan, B., Basith, S., Chakkarapani, S.K., Kang, S.H., Lee, G., Paik, M.J., Park, C.B., Silica-coated magnetic nanoparticles induce glucose metabolic dysfunction *in vitro via* the generation of reactive oxygen species. *Arch. Toxicol.*, 93, 1201–1212, 2019.
29. Puri, A., Nanoparticles: Crossing barriers and membrane interactions. *Mol. Membr. Biol.*, 27, 213–214, 2010.
30. Sharon, M., *Nanoparticles for Therapeutic Applications*, Wiley, Scrivener Publishing LLC, 2022, DOI:10.1002/9781119764205.
31. Pope, C.A., II, Burnett, R.T., Thun, M.J., *et al.*, Lung cancer, cardiopulmonary mortality, and long-term exposure to fine particulate air pollution. *JAMA*, 287, 1132–1141, 2002.
32. Dechsakulthorn, F., Hayes, A., Bakand, S., Joeng, L., Winder, C., *In vitro* cytotoxicity of selected nanoparticles using human skin fibroblasts. *Altern. Anim. Test. Exp.*, 14 397–400, 2008.
33. Foldbjerg, R., Dang, D.A., Autrup, H., Cytotoxicity and genotoxicity of silver nanoparticles in the human lung cancer cell line, A549. *Arch. Toxicol.*, 85, 743–750, 2011.
34. Kumah, E.A. and Fopa, R.D., Harati, S. *et al.*, Human and environmental impacts of nanoparticles: a scoping review of the current literature. *BMC Public Health*, 23, 1059, 2023, https://doi.org/10.1186/s12889-023-15958-4.
35. Harper, B.J., Engstrom, A.M., Harper, S.L., Mackiewicz, M.R., Impacts of differentially shaped silver nanoparticles with increasingly complex hydrophobic thiol surface coatings in small-scale laboratory microcosms. *Nanomaterials*, 14, 8, 654, 2024, DOI: 10.3390/nano14080654.
36. New NIOSH Document Highlights Safe Practices for Working with Nanomaterials, June 2012.
37. Biodegradation technology of organic and inorganic pollutants. *IntechOpen*, 2022, Available at: http://dx.doi.org/10.5772/intechopen.94650.
38. Additionally, the long-term effects of nanomaterials on soil microbiota and ecosystem functioning are not well understood.45 To address these research gaps and mitigate possible risks associated with the use of nanomaterials in agriculture, interdisciplinary research collaborations and regulatory frameworks are needed.
39. Arratia, F., Olivares-Ferretti, P., García-Rodríguez, A., Marcos, R., Carmona, E. R., Comparative toxic effects of copper-based nanoparticles and their microparticles in Daphnia magna by using natural freshwater media. *N. Z. J. Mar. Freshwater Res.*, 53, 3, 460–469, 2019.
40. Gupta, R. and Xie, H., Nanoparticles in daily life: applications, toxicity and regulations. *J. Environ. Pathol. Toxicol. Oncol.*, 37, 3, 209–230, 2018.

41. Isibor, P.O., Regulations and policy considerations for nanoparticle safety, in: *Environmental Nanotoxicology*, P.O. Isibor, G. Devi, A.A. Enuneku (Eds.), Springer, Cham, 2024, https://doi.org/10.1007/978-3-031-54154-4_14.
42. Karan, A., Nanotechnology in agriculture: enhancing crop productivity & sustainability. *J. Nanosci. Curr. Res.*, 9, 230, 2024.
43. Kay, L. and Shapira, P., The potential of nanotechnology for equitable economic development: the case of Brazil, in: *Nanotechnology and the Challenges of Equity, Equality and Development. Yearbook of Nanotechnology in Society*, vol. 2, S. Cozzens, and J. Wetmore (Eds.), Springer, Dordrecht, 2010, https://doi.org/10.1007/978-90-481-9615-9_19.
44. Whitman, J., *Nanotechnology and Dual-Use Dilemmas. On the Dual Uses of Science and Ethics: Principles, Practices, and Prospects*, B. Rappert and M.J. Selgelid (Eds.), pp. 13–28, ANU Press, JSTOR, 2013, http://www.jstor.org/stable/j.ctt5hgz15.7. Accessed 9 July 2024.
45. Hansson, S.O., How extreme is the precautionary principle? *Nanoethics*, 14, 245–257, 2020. https://doi.org/10.1007/s11569-020-00373-5.

10

Diseases and Virus Theranostics for Rewriting the Ethical Risk Management

Without equity, pandemic battles will fail. Viruses will simply recirculate, and perhaps undergo mutations or changes that render vaccines useless, passing through the unprotected populations of the planet.

Laurie Garrett

Abstract
The COVID-19 pandemic spotlighted the need for ethical management in theranostics—technologies combining diagnostics and therapy. This chapter discusses ethical challenges related to informed consent, data privacy, equitable access, and clinical trial integrity in developing personalized treatments. Theranostics, while promising, must be implemented with rigorous oversight to safeguard patient rights and public trust. Ethical decision-making in viral disease management requires balancing innovation with safety, transparency, and fairness in access to care.

Keywords: Virus, infectious, theranostic, cancer, neurological, risk-management, virus-management, ethical-management

10.1 Introduction

Management of an infectious disease, especially one caused by viruses, demands a combination of therapeutics and diagnostics. This is known as "theranostics," which offers an innovative approach to managing diseases and viruses by providing a more personalized and precise treatment strategy. The combination of therapeutic and diagnostic techniques holds great promise for the treatment of diseases, especially viral infections. However, it also raises several concerns that need to be addressed to ensure ethical and effective management.

Madhuri Sharon. Risk, Safety, and Ethical Concerns in Nanotechnology, (223–234) © 2025 Scrivener Publishing LLC

Figure 10.1 Series of some recorded viral pandemics across the globe.

History has recorded a series of viral pandemics that have occurred across the globe (Figure 10.1): Antonine Plague (165-180 AD; Plague of Justinian (541-549 AD); Black Death (1347-1351); Spanish Flu (1918-1919); Asian Flu (1957-1958), Hong Kong Flu (1968-1969); HIV/AIDS (1981-still continuing); H1N1 Influenza or Swine Flu (2009-2010); and now the COVID-19 Pandemic (2019-present). This is a cause for concern that should be utilized to inform ethical considerations for risk management.

Reliability and uncertainty related to ensuring the diagnostic accuracy of theranostic tools is crucial. Misdiagnosis or false positives/negatives can lead to inappropriate treatments and patient harm [1]. The long-term effects and efficacy of theranostic approaches are still being studied. Uncertainty in outcomes can pose risks to patients and complicate treatment decisions.

10.2 Theranostics in Disease and Virus Management

Theranostics in disease and virus management demands (i) personalized medicine, (ii) advanced diagnostic tools, and (iii) innovative technologies.

For personalized medicine there is a need for tailored treatments. Theranostics allows for treatments to be customized based on an individual's specific disease profile. For example, in cancer treatment, targeted therapies are developed based on the genetic makeup of a patient's tumor, along with real-time monitoring. Diagnostic tools that can continuously monitor the patient's response to therapy will allow for timely adjustments to treatment plans.

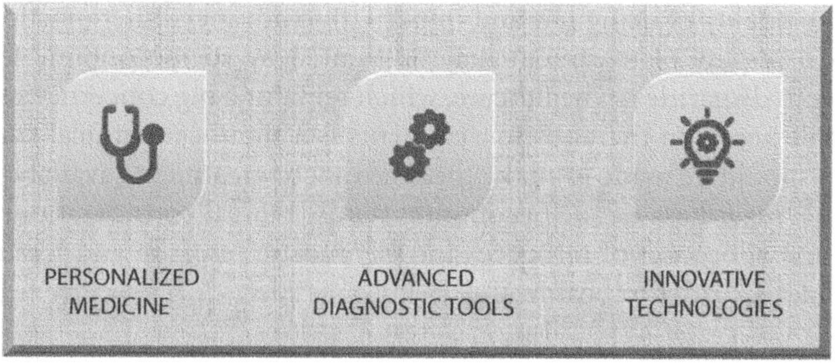

Figure 10.2 Theranostic applications for disease and virus management.

For such treatment, advanced diagnostic tools are required. Nowadays, molecular imaging techniques such as PET, MRI, and CT scans are used in combination with specific biomarkers to visualize disease progression and treatment response at the molecular level. Identifying specific biomarkers that indicate the presence and progression of a disease helps in early diagnosis and better monitoring of therapeutic outcomes.

However, rather than remaining stagnant with the available characterization tools; there is a need for innovative therapeutics, such as gene therapy, which modifies or replaces defective genes to treat genetic disorders or enhance the body's ability to fight diseases. Moreover, an innovative technique is also being practiced in nanomedicine using engineered nanoparticles to deliver drugs directly to the affected cells, thereby minimizing side effect and improving the efficacy of the treatment (Figure 10.2).

10.3 Ethical Risk Management and Risk Rewriting

Ethical considerations enjoin to practice informed consent. Patients must be fully informed about the nature of theranostic treatments, including potential risks and benefits, along with privacy. The collection and use of genetic information raise concerns about data privacy and potential misuse. Clear communication about the limitations and uncertainties of theranostic treatments is necessary to ensure patients make informed choices. Theranostics often involve the collection and analysis of sensitive patient data. Ensuring robust data protection measures is critical to safeguard patient privacy. Ethical guidelines must be established to govern the use of patient data in theranostic research and applications. Equity of access (i.e., the quality of being fair and impartial) to advanced theranostic treatments or equity of treatment may be limited by socioeconomic factors, leading to disparities in healthcare, which remains a big concern. Ensuring equitable access to theranostic treatments is a significant ethical concern. Efforts should be made to make these advanced treatments available to all patients, regardless of socioeconomic status. Ethical considerations must guide the allocation of resources for theranostic research and treatments to avoid disparities in healthcare.

10.3.1 Ethical Risk Management

The integration of theranostics into clinical practice may disrupt traditional diagnostic and treatment paradigms. Ethical guidelines should support clinicians in navigating these changes. Ensuring the safety of new theranostic technologies through rigorous clinical trials and ongoing monitoring has to be the most important criteria for risk management. There has to be proper regulation for this by establishing comprehensive regulatory frameworks to oversee the development and implementation of theranostic tools and therapies, as well as cost-effectiveness. The high costs of developing and implementing theranostic technologies must be balanced with their potential to reduce overall healthcare costs through more effective and efficient treatments. The principle of beneficence requires that theranostic treatments provide more benefits than harm. Careful risk assessment and management are necessary to uphold this principle of balancing benefits and risks. Nonmaleficence emphasizes the importance of avoiding harm to patients. This includes minimizing side effects and adverse reactions associated with theranostic treatments.

10.3.2 Rewriting Risk

For rewriting risk, predictive analytics is to be followed using data from diagnostic tools to predict disease outcomes and tailor preventative measures, followed by adaptive management. Treatment plans should be continuously updated based on real-time diagnostic information to mitigate risks associated with disease progression or treatment side effects. Educating or empowering patients about theranostic options actively involves them in their own healthcare decisions, potentially leading to better compliance and outcomes.

10.4 Some Notable Examples of Theranostic Approaches

Some examples of the practice of theranostics are mentioned below.

10.4.1 Theranostics for Cancer Treatment

Cancer theranostics, such as the use of radio-labeled antibodies for both imaging and treating tumors, ae used in oncology. Radiotracers used in locating cancer cells in the body are specifically designed for positron emission tomography (PET) imaging. Once identified, a therapeutic radiotracer is administered to target and destroy these cells. This approach is used in treating cancers like thyroid, neuroendocrine, and prostate cancers. Carbon dots [2] and gold nanoparticles can be used for both imaging and therapy. They enhance contrast in imaging techniques like CT scans and can be heated to destroy cancer cells through hyperthermia [3]. Magnetic nanoparticles are used as contrast agents in magnetic resonance imaging (MRI) for tumor detection. These nanoparticles can also deliver drugs directly to the tumor site, minimizing side effects. Superparamagnetic iron oxide nanoparticles (SPIONs) are used for MRI and can be functionalized to carry therapeutic agents, allowing simultaneous imaging and treatment [4]. Monoclonal antibodies are linked to radioactive isotopes or chemotherapeutic drugs; these antibodies target specific cancer cells, enabling precise imaging and targeted therapy [5]. In photoacoustic imaging and therapy, photoacoustic agents can absorb light and convert it into sound waves, providing high-resolution imaging used to deliver photothermal therapy, in which the absorbed light generates heat to destroy cancer cells [6].

10.4.2 Theranostics for Infectious Disease Caused by Virus

An infectious disease spreads from person to person. Nearly all viral infections are contagious, because viruses need to infect living things to reproduce, so human viruses survive by spreading between people. Ultrasound-triggered drug release using microbubbles as contrast agents can be loaded with drugs and injected into the bloodstream. When exposed to ultrasound waves, they release the drug at the targeted site, enhancing treatment efficacy. In the case of infectious diseases, developing rapid diagnostic tests combined with targeted antiviral or antibacterial therapies to combat diseases like HIV, hepatitis, and COVID-19 are being considered [7].

10.4.3 Theranostics for Neurological Disorders Caused by Virus

There are some neurological disorders that can be caused by various viral infections. The neurology of acute viral diseases encompasses [8]:

- *Herpes Simplex Virus (HSV)* can cause encephalitis, a severe brain infection leading to inflammation and damage.
- *Varicella-Zoster Virus (VZV)* reactivation can lead to shingles and, in some cases, encephalitis or myelitis (spinal cord inflammation).
- *Human Immunodeficiency Virus (HIV)* causes HIV-associated neurocognitive disorders (HAND) and opportunistic infections such as progressive multifocal leukoencephalopathy (PML). The HIV virus can cause a range of neurocognitive impairments, and the weakened immune system can allow other infections to affect the nervous system.
- *Enteroviruses* caused disorders are poliomyelitis (leading to paralysis) and viral meningitis.
- *Rabies Virus* causes severe encephalitis and is almost always fatal once symptoms appear.
- *West Nile Virus (WNV)* can lead to neurological complications such as encephalitis, meningitis, and paralysis.
- *Japanese Encephalitis Virus (JEV)* infection can cause severe brain inflammation, leading to neurological damage.
- *Zika Virus* causes congenital Zika syndrome and Guillain-Barré syndrome. It can cause microcephaly and other severe brain defects in fetuses, and it has been associated with Guillain-Barré syndrome in adults.
- *Influenza Virus* causes severe influenza infections that can lead to encephalitis and Reye's syndrome, which affects children and teenagers.
- *Measles Virus* causes subacute sclerosing panencephalitis (SSPE), a rare chronic form of progressive brain inflammation caused by a persistent infection with the measles virus.

These examples highlight how viral infections can lead to a range of neurological disorders, affecting the brain, spinal cord, and peripheral nerves.

For rewriting the risk and ethical management against diseases caused by virus; early diagnosis and treatment are crucial to manage viral neurological diseases effectively. This should include clinical assessment of history and symptoms along with comprehensive examination to assess cognitive function, motor skills, sensory response, and reflexes. Moreover, cerebrospinal fluid (CSF) analysis of white blood cells for testing specific viruses, and blood test to detect viral antigens, antibodies, and other markers of infection, should be done. Imaging studies like MRI and CT should be used to detect inflammation, lesions, and other abnormalities in the brain and spinal cord. Moreover, EEG (Electroencephalography) a non-invasive method for brain activity measurement and PCR (Polymerase Chain Reaction), and Serological detection reflects the severity of symptoms. These diagnostic methods will help in providing early treatment.

Many antiviral medications have already been developed, such as acyclovir or valacyclovir for herpes simplex virus; acyclovir, valacyclovir, or famciclovir for varicella-zoster virus; antiretroviral therapy (ART) to control HIV and prevent neurological complications; and oseltamivir or zanamivir for treating influenza. In the case of post-exposure prophylaxis for rabies, rabies immunoglobulin is administered, while in cases like encephalitis or myelitis, corticosteroids are often used to reduce inflammation. Moreover, supportive intensive care, hydration and nutrition are part of therapy, as well as rehabilitation to help recover motor skills and strength and assist in regaining daily living skills, which are also part of the theranostic approach.

To effectively manage viral neurological diseases and improve patient outcomes, preventive measures such as vaccinations, vector controls, public health measures, and safe sex practices are essential. And proper hygiene like handwashing and avoiding contact with contaminated food and water should be practiced to prevent the spread of enterovirus infections.

Theranostics holds great promise for revolutionizing disease and virus management by offering more personalized, effective, and adaptive treatments. However, careful consideration of ethical issues and robust risk management strategies are essential to maximize the benefits and minimize potential harms. Theranostics may change how diseases are diagnosed and treated, potentially leading to a redefinition of disease concepts. Ethical frameworks should address the implications of these changes.

10.4.4 Theranostics in Viral Neurological Diseases

Theranostics combines therapy and diagnostics to provide a more targeted and personalized approach to disease management. In the context of viral neurological diseases, theranostics plays a crucial roles in the treatment and management of viral neurological diseases. It can be applied in several ways:

(i) *Personalized Medicine:* This mainly incorporates genetic profiling, i.e., identifying genetic markers that predict the response to specific antiviral therapies, and biomarkers that monitor disease progression and treatment response.

(ii) *Nanotechnology-Based Diagnostics:* Nanoparticles are used to develop highly sensitive diagnostic tools for early detection of viral infections. For example, nanodiagnostics can identify viral fragments, specific proteins, and antibodies with high precision [9].

(iii) *Targeted Therapy:* Refers to the development and use of antiviral drugs that specifically target the virus causing the neurological condition; or gene therapy, which explores gene-editing technologies like CRISPR to target and inactivate viral genes in the nervous system. Targeted antiviral drug delivery can also be done using nanoparticles directly in infected cells, minimizing side effects and improving treatment efficacy. This targeted approach is particularly useful in treating diseases like COVID-19.

(iv) *Molecular Imaging and Real-Time Monitoring:* Molecular imaging utilizes imaging techniques to visualize viral activity within the brain, which allows monitoring the effectiveness of treatments [10]. Whereas, real-time monitoring employs wearable technology and mobile health applications to continuously monitor neurological functions and treatment efficacy. Another approach could be multimodal imaging, i.e., combining imaging techniques with therapeutic agents to allow real-time monitoring of treatment progress. This helps in adjusting therapies based on the patient's response [11].

10.5 Integration of Theranostics with Ethical Management

Although theranostics offers promising advancements in managing viral diseases, it also raises several ethical concerns that need to be addressed. Therefore, there is a need to combine theranostics and ethical management [12]. The first step in managing viral neurological diseases ethically involves informed consent and also addressing various issues.

(i) *Informed Consent:* Patients have to be educated to ensure that they understand the nature of their condition, the proposed treatments, and their potential risks and benefits; and must voluntarily consent to treatment and research studies without coercion. Transparency is crucial to ensure that patients can make informed decisions.

(ii) *Privacy, Confidentiality, and Data Security:* This is another area in which patient's consents should be taken, because theranostics often involves the collection of sensitive patient data. Ensuring robust data protection measures is essential to safeguard the privacy of patients, and for safeguarding their personal and medical information, especially in genetic and biomarker studies. It should be ensured that patient data be kept solely for the purposes consented to and is not misused. Personalized treatments and advanced diagnostics should be implemented with informed consent and respect for patient autonomy. Finally, the ethical implications of genetic profiling and targeted therapies should be addressed.

(iii) *Equitable Access:* Efforts should be made to ensure that advanced theranostic treatments and preventive measures are accessible to all patients, regardless of socioeconomic status. This includes addressing disparities in healthcare access, especially by actively working to reduce those disparities that affect vulnerable populations.

(iv) *Balanced Decision-Making:* Decisions should be made by weighing the benefits of advanced theranostic approaches against potential risks and ethical concerns; and engaging patients in shared decision-making that considers their values and preferences. The principle of beneficence requires that theranostic treatments provide more benefits than harm. Careful risk-benefit analysis, risk assessment and management are necessary to uphold this principle. There should be a continuous review of the ethical implications of new theranostic technologies, and guidelines should be adapted as necessary. Also, transparency and accountability in the development and application of theranostic tools should be promoted.

(v) *Ethical Management:* It is imperative to conduct clinical trials with rigorous ethical standards, including oversight by institutional review boards (IRBs). Along with this, the patient's safety has to be prioritized in experimental treatments by providing adequate monitoring and support. Establishing regulatory frameworks to oversee the development and application of theranostic technologies is crucial. This includes setting guidelines for safety, efficacy, and ethical considerations.

(vi) *International Collaboration:* This should be promoted to address viral outbreaks and neurological diseases, ensuring resource sharing and joint efforts. The ethical implications of global health policies, such as travel restrictions and vaccination mandates, should also be addressed.

(vii) *End-of-Life Care:* Appropriate palliative care should be provided for patients with terminal viral neurological

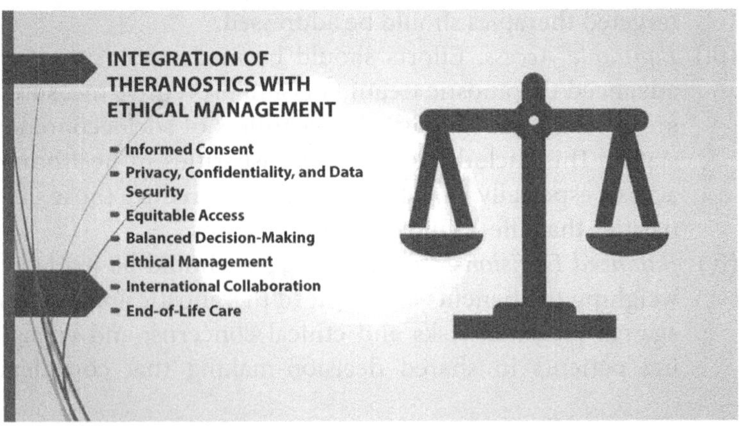

Figure 10.3 Some of the concerns in integrating theranostics with ethical management.

diseases; and advanced directives should be given respecting patients' wishes regarding end-of-life care and treatment withdrawal (Figure 10.3). By integrating theranostics and ethical management, healthcare providers can offer advanced, personalized care for viral neurological diseases while ensuring that ethical standards are upheld, and patients' rights are respected.

10.6 Concluding Remarks

History has recorded various virus pandemics across the globe in the past first century BC till the last decade of COVID. Management of infectious diseases caused by viruses demands a theranostics approach. Theranostics in virus-related disease management demands advanced diagnostic tools, innovative technologies, and personalized medicine. This chapter has touched upon these concerns through ethical management enjoined to practice informed consent and risk rewriting. In summary, while theranostics offers significant potential for managing viral diseases, it also presents ethical challenges that require careful consideration. By addressing these concerns, we can ensure that theranostic approaches are implemented responsibly and ethically, benefiting patients and society as a whole.

References

1. Jansen, S.N.G., Kamphorst, B.A., Mulder, B.C., *et al.*, Ethics of early detection of disease risk factors: A scoping review. *BMC Med. Ethics*, 25, 25, 2024. https://doi.org/10.1186/s12910- 024-01012-4.
2. Kwee, Y., Kristanti, A.N., Sharon, M., Fahmi, M.Z., A review of promising selected agents combined with carbon dots for biomedical applications. *Int. J. Nanosci. Nanotechnol.*, 18, 1, 11–44 11, March 2022.
3. Yasir, M., Mishra, R., Tripathi, A.S., Maurya, R.K., Shahi, A., Zaki, M.E.A., Al Hussain, S.A., Masand, V.H., Theranostics: a multifaceted approach utilizing nano-biomaterials. *Discov. Nano*, 19, 1, 2024. DOI 10.1186/s11671-024-03979-w.
4. Sharon, M., Oza, G., Gupta, A., Pandey, S., Chapter 13: Super-paramagnetic iron oxide nanoparticles (Spins) as nano-flotillas for hyperthermia: A paradigm for cancer theranostics, in: *Advanced Nanomaterials Synthesis, Properties, and Applications*, S. Thomas, N. Kalarikkal, A.M. Stephan, B. Raneesh, A.K. Haghi. (Eds.), pp. 302–328, Apple Academic Press, USA, Canada, 2014.

5. Okamoto, S., Shiga, T., Tamaki, N., Clinical perspectives of theranostics. *Molecules*, 26, 8, 2232, 2021, doi:10.3390/molecules26082232. ISSN 1420-3049. PMC 8070270. PMID 33924345.
6. Pini, C., Gelardi, F., Sollini, M., Present and future of target therapies and theranostics: refining traditions and exploring new frontiers—highlights from annals of Nuclear Medicine 2021. *Eur. J. Nucl. Med. Mol. Imaging*, 49, 11, 3613–3621, 2022, doi: 10.1007/s00259.
7. Kwee, Y., Zhou, Y., Fahmi, M.Z., Sharon, M., Kristanti, A.N., Review: Progress on applying carbon dots for inhibition of RNA virus infection. *Nanotheranostics*, 6, 4, 436–450, 2022, doi: 10.7150/ntno.73918.
8. Krett Jonathan, D., David Beckham, J., Tyler, K.L., Piquet, A.L., Chauhan, L., Wallace, C.J., Pastula, D.M., Kapadia, R.K., Neurology of acute viral infections. *Neurohospitalist.*, 12, 4, 632–646, 2022, doi: 10.1177/19418744221104778.
9. Sahana, R., Archunan, G., Sivakumar, M., Selvan, S.T., Lenin Fred, A., Kumar, S., Gulyás, B., Padmanabhan, P., Theranostic applications of nanoparticles in neurodegenerative disorders. *Int. J. Nanomed.*, 13, 5561–5576, 2018, doi: 10.2147/IJN.S149022.
10. Phelps, M.E., Positron emission tomography provides molecular imaging of biological processes. *PNAS*, 97, 16, 9226–9233, 2000, https://doi.org/10.1073/pnas.97.16.9226.
11. Shimojo, M., Ono, M., Takuwa, H., Mimura, K., Nagai, Y., Fujinaga, M., Kikuchi, T., Okada, M., Seki, C., Tokunaga, M., Maeda, J., Takado, Y., Takahashi, M., Minamihisamatsu, T., Zhang, M.-R., Tomita, Y., Suzuki, N., Maximov, A., Suhara, T., Minamimoto, T., Sahara, N., Higuchi, M., A genetically targeted reporter for PET imaging of deep neuronal circuits in mammalian brains. *EMBO J.*, 40, e107757, 2021, https://doi.org/10.15252/embj.202110775.
12. Pascual, T.N.B., Paez, D., Iagaru, A., Gnanasegaran, G., Lee, S.T., Sathekge, M., Buatti, J.M., Giammarile, F., Al-Ibraheem, A., Pardo, M.A., Baum, R.P., De Bari, B., Ben-Haim, S., Blay, J.-Y., Brink, A., Estrada-Lobato, E., Fanti, S., Golubic, A.T., Hatazawa, J., Israel, O., Kiess, A., Knoll, P., Louw, L., Mariani, G., Mirzaei, S., Orellana, P., Prior, J.O., Urbain, J.-L., Vichare, S., Vinjamuri, S., Virgolini, I., Scott, A.M., Guiding principles on the education and practice of theranostics. *Eur. J. Nucl. Med. Mol. Imaging*, 51, 2320–2331, 2024, https://doi.org/10.1007/s00259-024-06657-2.

11

COVID-19 Therapy and Vaccine: An Eye Opener to Risk and Ethics

Without equity, we cannot end COVID-19, HIV or any other pandemic.

Peter Sands, The Global Fund

Abstract
This chapter reviews ethical and safety concerns surrounding the accelerated development of COVID-19 vaccines. Despite rapid regulatory approvals, questions about efficacy, long-term effects, and adverse reactions—including myocarditis, clotting disorders, and anaphylaxis—remain. The novelty of mRNA and vector-based vaccines demands ongoing research. Public mistrust and vaccine hesitancy have highlighted the need for transparency and risk communication. Emerging detection technologies like SERS offer hope, but a balanced approach is critical to ensure public confidence and health protection.

***Keywords*:** Covid-19, SARS-CoV-2, therapies, adverse, vaccine, m-RNA, vector-based, protein-subunit

11.1 Introduction

In 2019 the world became engulfed by COVID-19, one of the most widespread and significant public health crises in decades, which has become among the leading causes of death globally. In this chapter, we are going to dive into the intriguing intersection of COVID-19 therapy, vaccines, and ethical considerations. It will address the scientific, ethical,tt and societal aspects of COVID-19 therapies and vaccines. COVID-19 is a severe acute respiratory syndrome caused by coronavirus 2 (SARS-CoV-2). The SARS-CoV-2 infection causes a hyperactive host immune response leading to an excessive inflammatory reaction with the release of many cytokines—a cytokine storm—commonly noticed in acute respiratory distress

Madhuri Sharon. Risk, Safety, and Ethical Concerns in Nanotechnology, (235–256) © 2025 Scrivener Publishing LLC

syndrome, sepsis and fulminant multiorgan failure. It is an infectious disease that spreads when an infected person breathes out, sneezes, coughs or speaks, which releases tiny droplets of respiratory fluid into the air, which can contain the virus. Other people can breathe in these droplets, or they can land on the eyes, nose, or mouth of others. People with COVID-19 have a wide range of symptoms ranging from mild symptoms to severe illness. Symptoms may appear 2–14 days after exposure to the virus, and may start as mild, and some people will progress to more severe symptoms.

11.1.1 An Overview of COVID-19 and History of SARS-CoV-2

In December 2019, a new coronavirus called severe acute respiratory syndrome coronavirus 2 (SARS-CoV-2) emerged in Wuhan, China [1] (Figure 11.1). It is a betacoronavirus that contains a single-stranded positive-sense RNA virus with a genome consisting of 29,903 nucleotides [2]. The SARS-CoV-2 virus causes Coronavirus disease 2019 (COVID-19). The World Health Organization (WHO) declared the COVID-19 outbreak a pandemic in 2020. Both WHO and the CDC issued recommendations for preventing and treating COVID-19, and by January 2021, COVID-19 was expected to have caused about 2.5 million deaths worldwide. Therefore, there was a sense of urgency in developing therapies and vaccines. This chapter serves as both a guide and a call to action. Let us focus not only on the science but also on the ethical compass that guides our pandemic response. A better understanding of coronaviruses and discoveries regarding their genetics and molecular biology have paved the way for the rapid

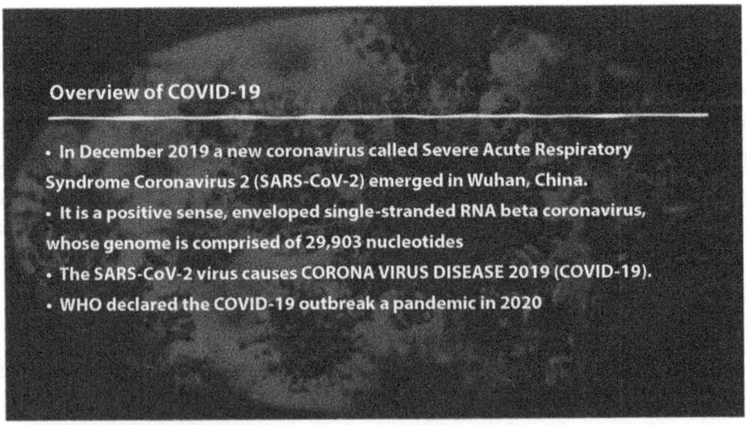

Figure 11.1 An overview of COVID-19 caused by the SARS-Co-2 virus, from its origin (in 1919) to the pandemic stage (in 2020).

development of virus-specific therapeutics and vaccines needed for the elimination of COVID-19.

The question is why is it crucial to discuss both the therapies and vaccines as well as risks and ethical considerations of COVID-19?

11.2 COVID-19 Therapies

Since the beginning of the pandemic, the scientific medical community has worked on therapeutic procedures that interfere with the exaggerated immune response. Thromboembolic complications are widespread in patients who are critically ill with COVID-19. Anticoagulant therapy was initially considered a cornerstone in hospitalized patients and even in the early post-discharge period; however, later trials have aborted the clinical benefits except for suspicion of or confirmed thrombosis.

Existing discoveries based on the basic science of coronaviruses across the globe has advanced the pace of developing specific drugs to target SARS-CoV-2; and a remarkable vaccine to prevent COVID-19 was also facilitated largely by these existing discoveries, as well as technological advances in molecular biology and genetics and nucleic acid-based vaccine development. Targeted therapies have been one of the first steps in treating COVID-19; and those being tested for their clinical efficacy include (Figure 11.2):

(i) Use of antibodies to prevent virus entry. The data on the effectiveness of existing targeted therapies from monoclonal antibodies against ACE2 are the best [3].
(ii) Nucleotide analogues to prevent viral replication are also accepted as one of the best targeted therapies.
(iii) Protease inhibitors prevent virion formation by blocking the enzyme that viruses need to replicate.
(iv) Antiviral drugs, repurposed medications, and supportive treatments: The antiviral agent remdesivir (the first drug approved by the FDA to treat COVID-19) [4], was effective in patients who were hospitalized with COVID-19 and had lower respiratory tract infection [5]. Izda *et al.* have also reported results of a number of clinical trials repurposing immunomodulatory agents, with positive results for dexamethasone in severe COVID-19.
(v) Ali [3] discussed the cardiovascular involvement in COVID-19. Such considerations have helped in developing therapies.

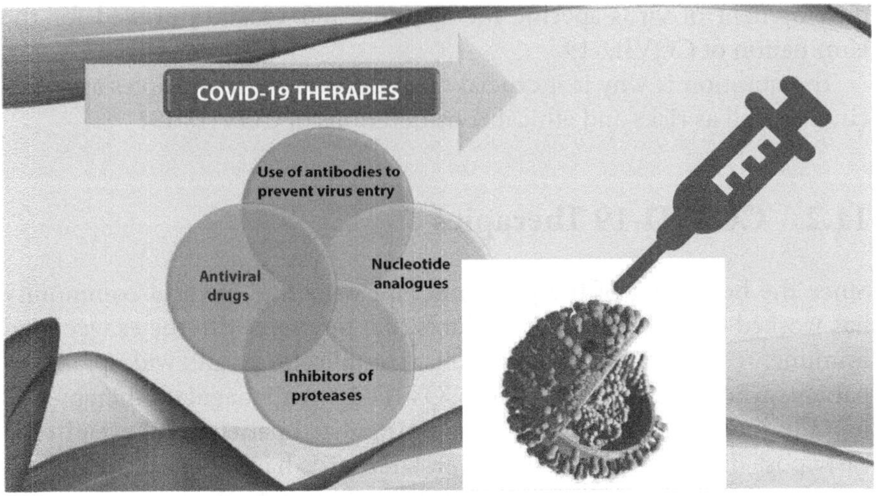

Figure 11.2 Various therapies in use to combat COVID-19.

11.2.1 Overview of Early Treatments

The fast spread of the COVID-19 pandemic across the globe required the development, testing and approval of a treatment at an unprecedented speed. Developing a treatment was challenging because of the complexities of COVID-19, which was due to variants, side effects, and limitations of available treatments, resulting in both successful and unsuccessful treatments, highlighting the risks involved.

Immunomodulatory therapies are still crucial in moderate to severe COVID-19. They include various medications, from steroids to hydroxychloroquine, tocilizumab and Anakinra. Anti-inflammatory agents, vitamin supplements, and antimicrobial therapy initially showed encouraging results, but there are limited data to review. Convalescent plasma, immunoglobulins, eculizumab, neutralizing IgG1 monoclonal antibodies and remdesivir have positively impacted inpatient mortality and hospital length of stay.

Initially many therapies were tried for treating SARS-CoV-2, most of them, such as convalescent plasma, hydroxychloroquine, colchicine, and enoxaparin, are no longer used [6]. Eventually, population-wide vaccination was proven to be the best tool to overcome the SARS- CoV-2 pandemic and help humanity return to regular life.

11.3 Vaccine Development and Science Behind COVID-19 Vaccine

The Operation Warp Speed Vaccine Initiative was a federal effort that supported multiple COVID-19 vaccine candidates to speed up the development, manufacture, and distribution of COVID-19 vaccines, therapeutics, and diagnostics. The phases of vaccine trials also focused on the challenges faced in rapid vaccine development. The COVID-19 vaccine is very good at preventing serious illness, hospitalization, and death. Across the globe, efforts were directed to develop different types of vaccine, such as ones based on mRNA, viral vector, protein subunit, etc. (Figure 11.3).

11.3.1 mRNA Vaccine

Even before the onset of COVID-19, in 1984 there were efforts by scientists at Harvard University to use a synthesized RNA enzyme to make biologically active messenger RNA (mRNA) in a lab. A similar process is still used today to make synthetic mRNA. Synthetic mRNA is used to study gene function and activity. In 1987, the Mayo Clinic mixed mRNA with fat droplets which, when human cells were added, absorbed the mRNA and made proteins. These experiments are considered early steps in the

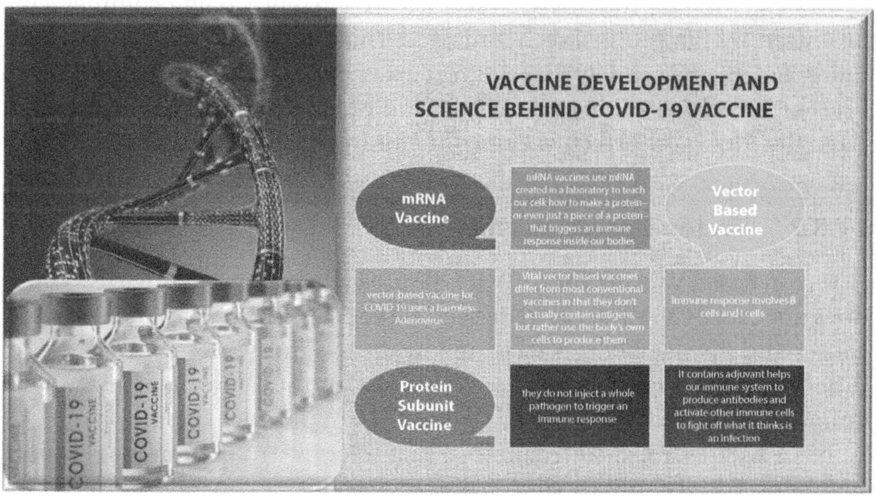

Figure 11.3 The science behind the development of three different types of COVID-19 vaccines that are now in use.

eventual development of mRNA-based COVID-19 vaccines. In the 1990s, mRNA was tested for treatment in rats and as an influenza and cancer vaccine in mice. In 2005, Kariko and Weissman's group [7] published a key discovery suggesting that mRNA could be altered and delivered effectively into the body to activate the body's protective immune system. This discovery paved the way to developing mRNA vaccine research, for which they received a Nobel Prize in 2023.

To trigger an immune response, instead of putting a weakened or inactivated germ into our bodies, mRNA vaccines use mRNA created in a laboratory to teach our cells how to make a protein—or even just a piece of a protein—that triggers an immune response inside our bodies. This immune response, which produces antibodies, is what helps protect us from getting sick from that germ in the future.

After vaccination, the mRNA enters the muscle cells and uses the cells' machinery to produce a harmless piece of spike protein. The spike protein is found on the surface of the virus that causes COVID-19. After the protein piece is made, our cells break down the mRNA and remove it, leaving the body as waste. Our cells display the spike protein piece on their surface, and our immune system recognizes that the protein does not belong there so they trigger our immune system to produce antibodies and activate other immune cells to fight off what they think is an infection. This is what your body might do if you got sick with COVID-19. This way, our bodies learn to help protect against future infection with the virus that causes COVID-19 The mRNA vaccines are currently being commercially produced by Pfizer BioNTech and Moderna. The FDA has also approved the Pfizer-BioNTech COVID-19 vaccine, now called COMIRNATY®, to prevent COVID-19 in people age 16 and older. The FDA has also authorized the Pfizer-BioNTech vaccine for children ages 5 through 15.

11.3.2 Vector-Based Vaccine

A vector-based vaccine for COVID-19 uses a harmless adenovirus (which does not cause COVID-19) to deliver the vaccine rather than the virus that causes the disease being vaccinated against. Viral vector-based vaccines differ from most conventional vaccines in that they don't actually contain antigens, but rather use the body's own cells to produce them.

It is a well-established technology that offers a strong immune response involving B cells and T cells. However, previous exposure to the vector could reduce effectiveness and it is a relatively complex procedure to manufacture. In 2021, the FDA gave emergency use authorization to the Janssen and Johnson & Johnson COVID-19 vaccine for vector-based vaccines.

Now it is manufactured by Janssen and Johnson & Johnson. CanSino Biologics developed Convidecia (Ad5-nCOV) using an Ad5 vector with genome of S protein of SARS-CoV-2; it is a single-dose vaccine.

There are discussions about gene therapies and COVID-19 vaccines in relation to viral vector-based approaches. Aledo-Serrano *et al.* [8] have raised a question about the potential interference between gene therapy and virus-based vaccines, which deserves careful consideration and a discussion involving the patient community.

11.3.3 Protein Subunit Vaccine

A protein subunit vaccine is a vaccine that uses fragments of a COVID-19-causing virus to trigger an immune response. These virus pieces are the spike protein. These vaccines are sometimes called acellular vaccines. Since, the fragments in a protein subunit vaccine are unable to cause disease, they are considered very safe. Protein subunit vaccines differ from other vaccines because they do not inject a whole pathogen to trigger an immune response. The vaccine also contains another ingredient called an adjuvant that helps the immune system respond to that spike protein in the future. The adjuvant helps our immune system to produce antibodies and activate other immune cells to fight off what it thinks is an infection. This vaccine is suitable for those who cannot receive live vaccines. Once our immune system recognizes the S proteins, it creates antibodies and defensive white blood cells, and if one of them later becomes infected with the COVID-19 virus, the antibodies will fight the virus. The Novavax COVID-19 vaccine is a protein subunit vaccine.

The pace at which vaccine development to prevent COVID-19 has taken place is quite remarkable. This is mostly due to a sense of urgency, as COVID quickly became a pandemic, and the fact that its development was facilitated largely by the existing technological advances in molecular biology and genetics and nucleic acid-based vaccine. The COVID-19 vaccine lessens the severity of the disease by teaching the immune system to recognize and fight the virus that causes it. COVID-19 vaccines approved by the US Food and Drug Administration (FDA) are safe and recommended for people who are pregnant or lactating, as well as for those intending to become pregnant. It usually takes about two weeks for the vaccine to become effective. However, no concrete evidence exists on the efficacy of the potential vaccines.

Mariarosaria Boccellino [9] has critically reviewed conflicting reports regarding the impact of COVID-19 vaccines on the immune system, especially their potential to induce the production of autoantibodies. She has suggested that in the fight against COVID-19 there are reliable and efficient virus identification methods. One innovative approach of key importance is surface-enhanced Raman spectroscopy (SERS), which offers the potential for rapid detection of multiple respiratory viruses. Her review of many articles contributes valuable insights into vaccine responses, treatment approaches, safety considerations, and virus detection methods, enhancing our understanding and strategies in the ongoing battle against COVID-19.

The important thing to know is that none of these vaccines can give you COVID-19. Vaccines do not use any live virus. They cannot cause infection with the virus that causes COVID-19 or other viruses. Moreover, COVID-19 vaccines do not affect or interact with our DNA. These vaccines do not enter the nucleus of the cell where our DNA (genetic material) is located, so they cannot change or influence our genes.

11.4 Potential Harm and Risk Assessment of COVID-19 Therapies and Vaccines

In today's scenario there is no need to highlight the benefits of vaccination in preventing severe disease and death. The rapid development and deployment of COVID-19 therapies and vaccines were critical in addressing the global health crisis. However, these efforts brought significant ethical, medical, and societal considerations. The potential harms and risks associated with COVID-19 therapies and vaccines must be carefully assessed to ensure that the benefits outweigh the risks and that the public is adequately informed. Though there have been extensive clinical trials and real-world data supporting vaccine safety for many diseases prior to applying them for efficacy and safety's sake, data on vaccine effectiveness, side effects, and the balance between benefits and risks are a must to consider. For example, for myocarditis (inflammation of the heart muscle), there was a need to address the evidence linking mRNA vaccines (Pfizer-BioNTech and Moderna) to myocarditis. Other expected potential harms were for conditions like Guillain-Barré syndrome, Bell's palsy, thrombosis with thrombocytopenia syndrome (TTS), and ischemic stroke. While some associations were found to exist, the overall risk remains low. There are limited data about the Janssen vaccine and its potential risks to TTS and Guillain-Barré syndrome.

To understand the risks involved in vaccines, an analysis of Risk vs. Benefit is important. The importance of transparent communication about the risks associated with therapies and vaccines is crucial, along with how benefits were justified in the face of uncertainty, and the role of public trust and consideration of potential long-term effects of both the disease and its treatments.

11.4.1 Potential Harmful Effects of COVID-19 Therapies

The harmful effects that may be expected due to different therapeutic approaches used for treating COVID-19 are shown in Figure 11.4 and discussed below.

11.4.1.1 Adverse Drug Reactions

The following adverse drug reactions are possible: (i) Immediate reactions may be experienced by some individuals who receive COVID-19 therapy drugs, such as allergic reactions, anaphylaxis, or other side effects. (ii) Long-term effects, which are still under investigation. There is a risk that treatments may lead to unanticipated long-term health consequences. (iii) Development of resistance to drug. Widespread use of antiviral drugs can lead to the development of resistance, reducing the effectiveness of these drugs in future cases.

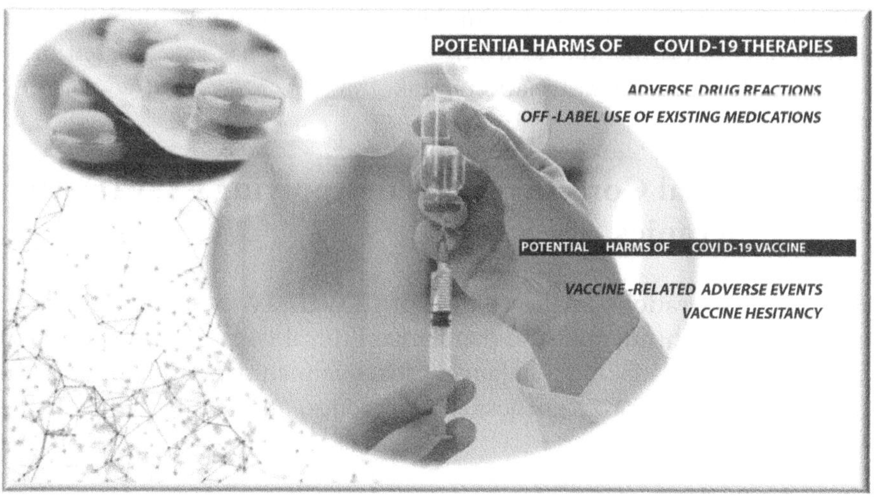

Figure 11.4 Harmful effects that may be caused by the use of different therapeutic approaches for treating COVID-19.

11.4.1.2 Off-Label Use of Existing Medications

The use of existing off-label medications may be inappropriate for COVID-19 treatment, leading to incorrect dosing that can result in toxicity or reduced efficacy. In addition to possibly being ineffective, some of these off-label drugs for COVID-19 can potentially provide a false sense of security and delay the use of more effective interventions.

11.4.2 Potential Harmful Effects of COVID-19 Vaccines

11.4.2.1 Vaccine-Related Adverse Events

Common side effects caused by vaccines are pain at the injection site, fatigue, fever, and muscle aches. There are rare chances that some individuals may exhibit severe adverse reactions, including myocarditis, blood clotting disorders, or anaphylaxis. Given the novel nature of mRNA and vector-based vaccines, there is a chance of unknown long-term effects of these vaccines. Ongoing research into their long-term effects on health by various groups may provide an answer to this.

11.4.2.2 Vaccine Hesitancy

The main reasons for hesitancy are uncertainty, public distrust, and misinformation. The rapid development and emergency use authorization of vaccines have led to public skepticism and vaccine hesitancy, potentially reducing vaccination rates and hindering herd immunity. Moreover, the spread of misinformation regarding vaccine safety has exacerbated hesitancy and led to lower vaccination uptake.

11.5 Ethical Considerations in Applying COVID-19 Therapies

Ethical considerations in the context of COVID-19 therapies have been a critical topic throughout the pandemic. Ethical considerations in these therapies involve striking a balance, proactive planning, and adhering to principles that prioritize fairness, transparency, and equity. Some of the key points to consider are:

11.5.1 Resource Allocation and Fairness

When healthcare resources are strained, such as during the pandemic, balancing fairness and equity becomes crucial. During the pandemic, healthcare workers faced extraordinary patient loads and had to make difficult decisions regarding treatment and vaccine distribution. States and institutions grappled with allocating scarce resources for treatments. Some promising practices emerged such as (i) Tailoring approaches: Striking a balance between context-specific approaches and consistent responses to avoid unfairness. (ii) Proactive planning: Setting up frameworks and expectations ahead of time to guide decision-making during scarcity. (iii) Public engagement: In Minnesota, for instance, public engagement was used to establish agreed-upon goals based on fairness and equity.

Myers *et al.* [10] have presented ethical arguments both for and against a mandate for healthcare workers to receive a vaccination against COVID-19. They argued that vaccine mandates rests on the ethical requirements for healthcare workers to place the well-being of the community as a whole above their own individual interests. Whereas, the key arguments against mandates state that an individual's autonomy must be primary, and that autonomy should not be overridden or ignored in the name of a "higher" good.

11.5.2 Ethical Guidance During COVID-19

Psychologists have emphasized the importance of ethics in practice, research, teaching, and applied work during the pandemic. Critical thinking and flexible decision-making were anchored in ethical principles. The World Health Organization (WHO) provided practical guidance on ethical considerations in COVID-19 research, emphasizing values like transparency and resource allocation (https://www.who.int/teams/health-ethics- governance/diseases/covid-19).

11.6 Ethical Considerations in Using Vaccine

Persad and Emanuel [11] have questioned whether it is ethical to offer incentives for getting vaccinated against COVID-19 (i.e., entry into million dollar lottery for adults; lottery ticket for state college tuition, room, board, and more for teens; some states in the USA offered gift cards; employees offered business of various types).

In their analysis, they discussed advantages of and objection to COVID-19 vaccine incentives. According to them, the main benefits are prevention of harm from COVID-19 and protection of a disadvantage population. Whereas, the objections were coercion, exploitation and distorted decision-making, which harm those already vaccinated, make vaccine look riskier; destroy public willingness to be vaccinated, and waste public funds.

Malekzadeh *et al.* [12] analyzed the ethical considerations of COVID-19 vaccination, and what lessons for the future we have learned. They have recommended that health system managers should be adequately prepared to solve the ethical problems posed by COVID-19 vaccination. Haste should be avoided when developing vaccination and more attention paid to safety standards, providing sufficient resources for a comprehensive vaccine supply, and close attention should also be paid to collective interests versus individual interests in order to meet community needs. The areas of ethical consideration should involve:

11.6.1 COVID-19 Vaccine Allocation Principles

The Advisory Committee on Immunization Practices (ACIP) in the United States considered four ethical principles for vaccine allocation: (i) Maximizing benefits while minimizing harms, (ii) Promoting justice, (iii) Mitigating health inequities, and. (iv) Ensuring transparency.

The allocation and distribution of COVID-19 vaccines varied significantly across different countries. Many countries implemented phased distribution plans that prioritized specific groups based on risk factors. These typically included high-risk individuals such as the elderly and those with underlying health conditions; healthcare workers exposed to the virus; and frontline workers, including essential workers in critical sectors. These phased approaches aimed to ensure that those most vulnerable or essential received vaccines first.

Some countries, like the United Arab Emirates (UAE) and Israel, managed vaccine distribution through centralized governance bodies. This approach allowed for better coordination and synchronization. However, other countries had state-by-state variations in vaccine eligibility, prioritization, and distribution. This lack of uniformity made it challenging to predict, replicate, and leverage gains from synchronicity.

High-income countries (HICs) often entered into direct contractual arrangements with vaccine suppliers, bypassing the COVAX mechanism. COVAX is a global initiative aimed at ensuring equitable vaccine distribution.

However, pharmaceutical companies sometimes failed to meet their contractual obligations with COVAX, affecting supply and predictability. COVAX allocated vaccines across 138 countries based on a framework developed by an expert group. This framework considered ethical, scientific, and public health factors. The World Health Organization (WHO) played a key role in vetting and guiding the allocation process.

11.6.2 Informed Consent

Informed consent during the COVID-19 pandemic, particularly in the context of emergency use authorizations (EUAs), has been a critical topic. To handle it, an EUA was employed. The EUA authority allowed the US FDA to facilitate the availability and use of medical countermeasures (MCMs) during public health emergencies, including infectious diseases like COVID-191. When the Secretary of Health and Human Services (HHS) declares that an EUA is appropriate, the FDA can authorize unapproved medical products or unapproved uses of approved medical products for emergency use. This authorization applies to diagnosing, treating, or preventing serious or life-threatening diseases caused by chemical, biological, radiological, or nuclear (CBRN) threat agents. In the context of EUAs, informed consent remains essential. However, there are unique challenges, such as if rapid development and deployment of COVID-19 vaccines and treatments is urgent and requires swift action, potentially affecting the traditional informed consent process. Moreover, limited data during EUAs must have been due to limited expedited trials and emergency circumstances. Hence, balancing the risk-benefit communication with urgency of vaccination with transparent communication would have been very difficult to maintain. Alternative options are that the participants should be informed about any available alternatives, even if they are not FDA approved.

Institutional Review Boards (IRBs) played a crucial role in reviewing consent processes during the pandemic by ongoing monitoring to ensure participants understand the evolving nature of knowledge about the vaccine or treatment. Researchers and healthcare providers adapted by enhancing communication so as to provide clear, concise information about the vaccine or treatment. Documenting consent was another step in ensuring proper documentation of informed consent, even in urgent situations. The Public Readiness and Emergency Preparedness Act (PREP Act) was applied to provide liability immunity for activities related to medical countermeasures against COVID-19. Declarations under the PREP Act support EUAs and address liability concerns.

In summary, informed consent during the COVID-19 pandemic required balancing urgency, transparency, and ethical oversight. Researchers and healthcare providers adapted to ensure participants were well-informed despite the unique challenges posed by EUAs.

11.6.3 Equitable Distribution/Access

The ethical issues surrounding the distribution of vaccines were prioritization, global equity, and access disparities. There is a need to explore the ethical challenges of vaccine allocation ensuring fair access across populations, both globally and within countries. Vaccine allocation strategies range from centralized governance to decentralized state-level decisions. Ethical considerations, risk-based prioritization, and global cooperation were essential in this unprecedented effort to immunize the world against COVID-19. Equitable access will provide global distribution of vaccines and therapies, particularly in low- and middle-income countries, to prevent unequal health outcomes. Justice and fairness in disparities in vaccine distribution and access are to be addressed, highlighting the importance of transparent communication about risks and benefits.

In 2020, it was reported by McClung *et al.* [13] in *Morbidity and Mortality Weekly Report* that "During the period when the U.S. supply of COVID-19 vaccines is limited, the Advisory Committee on Immunization Practices (ACIP) will make vaccine allocation recommendations. In addition to scientific data and implementation feasibility, four ethical principles will assist ACIP in formulating recommendations for the initial allocation of COVID-19 vaccine: 1) maximizing benefits and minimizing harms; 2) promoting justice; 3) mitigating health inequities; and 4) promoting transparency. So that ethical principles will aid ACIP in making vaccine allocation recommendations and state, tribal, local, and territorial public health authorities in developing vaccine implementation strategies based on ACIP's recommendations."

11.6.4 Ethics of Balancing Vaccine Mandates with Public Health

The ethics of balancing individual rights with public health requires maximizing benefits and minimizing harms to balance between individual protection and maintain community benefit. Since it is our duty to protect key populations, to achieve public health goals and protect those who are unable to protect themselves, eligible individuals should receive the COVID-19 vaccination. Without a mandate, these select populations would be at risk of serious harm due to their inability to receive the vaccine. An analysis of participants' experiences has led to the extraction of five main categories of themes and fifteen subcategories of the ethical considerations of COVID-19 vaccination: Safe and standard vaccine production, vaccine supply, fairness, respect for autonomy, and accountability were the main categories.

11.6.5 Transparent Communication

Transparent communication will be possible only if the public is educated about transparent and accurate information regarding the benefits and risks of COVID-19 therapies and vaccines.

To address misinformation, proactively combating misinformation through fact-based campaigns and engagement with trusted community leaders will be effective.

11.6.6 Data Privacy

Data privacy during the COVID-19 pandemic has been a crucial topic, especially as governments, institutions, and companies grapple with balancing public health measures and individual privacy rights. Data privacy is a concern related to the collection and sharing of personal health data during the pandemic. Following are the ways in which it was handled.

11.6.6.1 The Dilemma: Public Health vs. Personal Privacy

During the pandemic, authorities implemented various measures to control the spread of the virus. These measures often impacted personal freedom in three key areas: (i) Movement restrictions for physical distancing, public gathering limitations, quarantine, isolation, and lockdowns. (ii) Health reporting, including COVID-19 testing, temperature checks, health surveys (both public and private), and reporting by authorities and corporations. (iii) Manual and automated health tracking using contact tracing (including mobile apps) and proximity detection systems. However, some of these measures could potentially encroach upon individuals' privacy rights.

11.6.6.2 Stricter Data Protection Regulations

The European General Data Protection Regulation (GDPR) introduced in 2018, significantly raised awareness about data privacy. It set new standards for protecting the "rights and freedoms of data subjects" globally. Companies and institutions now face stricter rules regarding privacy and data protection. The possible "ePrivacy Regulation" further emphasizes privacy considerations.

11.6.6.3 Balancing Privacy and Health Protection

Companies and governments must navigate this delicate balance. Here, there were three productive actions for handling data privacy during the pandemic: (i) Thoughtfully consider privacy and cybersecurity implications when implementing pandemic-related measures. (ii) Be transparent about data collection, usage, and storage. Clear communication builds trust. (iii) Ensure that privacy standards are maintained even in exceptional circumstances.

11.6.6.4 Legal Guidance and Privacy Laws

Privacy laws still apply during public health crises. However, they are not barriers to appropriate information sharing. Authorities should strike a balance between public health needs and individual privacy rights while handling data during the pandemic. COVID-19 hasn't fundamentally changed data privacy concerns but has created a new context for data governance.

Solutions should address privacy and ethics while enabling effective pandemic response, because the goal is to protect public health without compromising personal privacy.

11.6.7 Risk Mitigation Strategies

Phase III trials are rigorous clinical trials. Therapies and vaccines should undergo Phase III clinical trials to establish safety and efficacy before widespread use; along with implementing robust post-market surveillance systems to monitor and address any emerging adverse events.

11.6.7.1 Some Case Studies in Ethics

Ethical dilemmas in conducting trials in developing countries or among vulnerable groups have posed special consideration. There is a need for granting emergency use and authorization related to ethical implications for vaccines and treatments without long-term data. Often misinformation has impacted public perception and trust in COVID-19 therapies and vaccines. Such cases help in learning lessons and getting insights on how future pandemics should be handled in terms of therapy and vaccine development with a focus on ethics, so that policies can be made accordingly. And they also emphasize the importance of global collaboration in addressing both risks and ethical challenges and how transparent communication can mitigate risks and build public trust.

The world has realized the need to balance rapid development with ethical responsibility. The lasting impact of COVID-19 on medical ethics and the future of vaccine and therapy development is the realization that equitable vaccine distribution needs to be ensured worldwide.

11.7 Role of Nanotechnology in COVID Therapy and Vaccine

Nanotechnology has played a significant role in both the therapy and vaccine development for COVID-19. Its unique properties allowed for improved drug delivery, diagnostic tools, and vaccine formulations. Nanotechnology has provided innovative solutions across the entire spectrum of COVID-19 management—from vaccine development and drug delivery to diagnostics and personal protective equipment (PPE). Its ability to manipulate matter at the nanoscale enables highly targeted therapies, improved vaccine delivery, and more efficient antiviral interventions, significantly contributing to the global response against COVID-19.

Below is a breakdown of the key roles of nanotechnology in combating COVID-19:

11.7.1 Nanotechnology in COVID-19 Vaccines

For mRNA delivery, nanotechnology was instrumental in developing mRNA vaccines like Pfizer-BioNTech (BNT162b2) and Moderna (mRNA-1273). These vaccines used lipid nanoparticles (LNPs) to deliver mRNA safely into the cells, allowing them to instruct cells to produce the spike protein of SARS-CoV-2, thus generating an immune response.

Lipid nanoparticles (LNPs) serve as carriers that protect the mRNA from degradation, facilitating its entry into the cells, where the body can then produce the desired immune response. They are key for stability, delivery, and enhancing the efficacy of the vaccines.

For targeted immunity, nanoparticles can be designed to target specific tissues or cells, helping the vaccines focus on the right areas of the immune system to trigger strong, specific immune responses [14].

11.7.2 Nanotechnology in COVID-19 Therapy

For delivery of antiviral drug, nanoparticles have been used to improve the delivery of drugs like remdesivir. Nanocarriers, such as liposomes, polymeric nanoparticles, and dendrimers, enhance the solubility, stability, and bioavailability of drugs, thus improving their efficacy. Nanotechnology allows for the controlled release of therapeutic agents over time, ensuring sustained drug delivery and reducing the frequency of dosing.

Nanomaterials like silver nanoparticles and carbon-based nanoparticles exhibit intrinsic antiviral properties. These materials are being explored for their ability to disrupt viral replication and viral envelope integrity [15].

11.7.3 Nanotechnology in COVID-19 Diagnostics

Nanoparticles have enabled faster, more accurate COVID-19 diagnostic tests. For example, gold nanoparticles (AuNPs) have been employed in rapid antigen tests for detecting SARS-CoV-2 by binding to viral proteins, leading to color changes that indicate the presence of the virus. Moreover, nanotechnology has facilitated the development of biosensors that detect viral RNA or proteins with high sensitivity and specificity. These biosensors use nanomaterials to enhance signal detection and improve diagnostic accuracy [16].

11.7.4 Nanotechnology in Immune Modulation for COVID-19

Some nanomaterials have been used to modulate the immune response in patients with severe COVID-19, particularly those suffering from cytokine storms. For example, nanoparticles can be designed to deliver anti-inflammatory agents or cytokine inhibitors to dampen excessive immune reactions [17].

11.7.5 Nanotechnology in Antiviral Personal Protective Equipment

Nanotechnology has improved the efficacy of personal protective equipment (PPE) like masks and gowns. Nanofiber filters enhance the filtration efficiency of masks, while antiviral coatings can be applied to surfaces to reduce the survival of the virus on them [18].

11.8 Conclusion

The development and deployment of COVID-19 therapies and vaccines involve significant risks and potential harms. A careful, ethically sound approach is essential to mitigate these risks while maximizing the benefits of these interventions. Ongoing research, transparent communication, and global cooperation are key to managing the challenges posed by the pandemic and ensuring public trust in the solutions offered. This requires ensuring transparency and addressing key risks.

References

1. Lu, R., Zhao, X., J. Li, P., Yang, B., Wu, H., Wang, W., Song, H., Huang, B., Zhu, N., Bi, Y., Ma, X., Zhan, F., Wang, L., Hu, T., Zhou, H., Hu, Z., Zhou, W., Zhao, L., Chen, J., Meng, Y., Wang, J., Lin, Y., Yuan, J., Xie, Z., Ma, J., Liu, W.J., Wang, D., Xu, W., Holmes, E.C., Gao, G.F., Wu, G., Chen, W., Shi, W., Tan, W., Genomic characterisation and epidemiology of 2019 novel coronavirus: implications for virus origins and receptor binding. *Lancet*, 395, 565–574, 2020.
2. Wu, F., Zhao, S., Yu, B., Chen, Y.M., Wang, W., Song, Z.G., *et al.*, A new coronavirus associated with human respiratory disease in China. *Nature*, 579, 265–269, 2020. [PMC free article] [PubMed] [Google Scholar].
3. A.J. and Marian, Current state of vaccine development and targeted therapies for COVID-19: Impact of basic science discoveries. *Cardiovasc. Pathol.*, 50, 107278, 2021, doi: 10.1016/j.carpath.2020.107278.
4. Adibzadeh, S., Amiri, S., Nia, G.E., Taleghani, M.R., Bijarpas, Z.K., Maserat, N., Maali, A., Azad, M., Behzad-Behbahani, A., Therapeutic approaches and vaccination in fighting COVID-19 infections: A review. *Gene Rep.*, 27, 101619, 2022, https://doi.org/10.1016/j.genrep.2022.101619.
5. Vladislav, I., Jeffries, M.A., Sawalha, A.H., COVID-19: A review of therapeutic strategies and vaccine candidates. *Clin. Immunol.*, 222, 108634, January 2021, https://doi.org/10.1016/j.clim.2020.108634.
6. Siddharth, C., Rodriguez-Guerra, M., Shaban, M., Pandey, N., Jaquez-Duran, M., Vittorio, T.J., COVID-19 therapy and vaccination: A clinical narrative review. *Drugs Context*, 12, 2022-7-2, 2023. doi: 10.7573/dic.2022-7-2; PMCID: PMC9914077, PMID: 36793450.
7. Karikó, K., Buckstein, M., Ni, H., Weissman, D., Suppression of RNA recognition by Toll- like receptors: the impact of nucleoside modification and

the evolutionary origin of RNA. *Immunity.*, 23, 2, 165–75, 2005 Aug. doi: 10.1016/j.immuni.2005.06.008. PMID: 16111635.

8. Aledo-Serrano, A., Gil-Nagel, A., Isla, J., Mingorance, A., Mendez-Hermida, F., Hernandez-Alcoceba, R., Gene therapies and COVID-19 vaccines: A necessary discussion in relation with viral vector-based approaches. *Orphanet J. Rare Dis.*, 16, 316, 2021, doi: 10.1186/s13023-021-01958-3, PMCID: PMC8284696 PMID: 34271959.
9. Boccellino, M., COVID-19 Pandemic: Therapeutic Strategies and Vaccines. *Int. J. Mol. Sci.*, 25, 1, 556, 2024, doi: 10.3390/ijms25010556 MCID: PMC10778581, PMID: 38203727.
10. Myers, M., Dunikoski, L., Brantner, R., Fletcher, D., Saltzberg, E.E., Urdaneta, A.E., Wedro, B., Giwa, A., An ethical anaylsis of the arguments both for and against COVID-19 vaccine mandates for healthcare workers. *J. Emerg. Med.*, 64, 2, 246–250, 2023, doi: 10.1016/j.jemermed.2022.11.005. PMCID: PMC9659509 PMID: 36746692.
11. Govind, P. and Ezekiel J., E., Ethical considerations of offering benefits to COVID-19 vaccine. *JAMA*, 326, 3, 221–222, 2021, doi: 10.1001/jama.2021.11045.
12. Malekzadeh, R., Abedi, G., Ziapour, A., Yıldırım, M., Amirkhanlou, A., Analysis of ethical considerations of COVID 19 vaccination: Lessons for future. *BMC Med. Ethics*, 24, 91, 2023, doi: 10.1186/s12910-023-00969-y. PMCID: PMC10612281 PMID: 37891543.
13. McClung, N., Chamberland, M., Kinlaw, K., Matthew, D.B., Wallace, M., Bell, B.P., Lee, G.M., Talbot, H.K., Romero, M.J.R., Oliver, S.E., Dooling, K., The Advisory Committee on Immunization Practices' Ethical Principles for Allocating Initial Supplies of COVID-19 Vaccine — United States, 2020. *Morbidity Mortality Wkly Rep.*, 69, 47, November 27, 2020.
14. Kumar, P.P., Khurana, N., Vyas, M., Sharma, V., Batiha, G.E.-.S., Kaur, H., Singh, J., Kumar, D., Sharma, N., Kaushik, A., Kumar, R., Aspects of nanotechnology for COVID-19 vaccine development and its delivery applications. *Pharmaceutics*, 15, 2, 451, 2023, doi: 10.3390/pharmaceutics15020451 PMCID: PMC9960567PMID: 36839773.
15. Gaurav, C., Marc, J., Madou, Kalra, S., Chopra, V., Ghosh, D., Martinez-Chapa, S.O., Nanotechnology for COVID-19: Therapeutics and vaccine research. *ACS Nano*, 14, 7760–7782, 2020.
16. Ruiz-Hitzky, E., Darder, M., Wicklein, B., Ruiz-Garcia, C., Martín-Sampedro, R., del Real, G., Aranda, P., Nanotechnology responses to COVID-19. *Adv. Healthc. Mater.*, 9, 2000979, 2020. 2000979 (1 of 26) © 2020 The Authors. Published by Wiley-VCH GmbH.

17. Weiss, C., Carriere, M., Fusco, L., Capua, I., Regla-Nava, J.A., Pasquali, M., Scott, J.A., Vitale, F., Unal, M.A., Mattevi, C., Bedognetti, D., Merkoçi, A., Tasciotti, E., Yilmazer, A., Gogotsi, Y., Stellacci Lucia, F,., Delogu, G., Toward nanotechnology-enabled approaches against the covid-19 pandemic. *ACS Nano*, 6, 6383–6406, 2020, https://doi.org/10.1021/acsnano.0c03697.
18. Phuna, Z.X., Panda, B.P., Hawala Shivashekaregowda, N.K., Madhavan, P., Nanoprotection from SARS-COV-2: would nanotechnology help in personal protection equipment (PPE) to control the transmission of COVID-19? *Int. J. Environ. Health Res.*, 33, 7, 670–699, 2023, doi: 10.1080/09603123.2022.2046710. Epub 2022 Mar 5. PMID: 35253535.

12

Social Ethics, Issues, and Anthropological Questions and Uncertainties in ID Detection

Ethics is knowing the difference between what you have a right to do and what is right to do.

Potter Stewart

Abstract
Focusing on social ethics, this chapter examines the moral principles guiding nanotechnology's integration into society. Ethical concerns include risk communication, transparency, autonomy, fairness, and the displacement of labor due to monopolization of knowledge. Regulatory practices vary globally, highlighting the need for national and regional oversight. Anthropology offers critical insight into cultural contexts, helping to identify ethical concerns across diverse communities. Social ethics ensures that nanotech development respects cultural values while upholding universal ethical standards.

Keywords: Social-ethics, anthropological, ID-detection, nanotoxicity, nanotechnology, safety-risk, ecological-risk

12.1 Introduction

Social ethics and anthropology are related concerns. Anthropology is the study of humanity, including human behavior, human biology, cultures, and societies; and has been concerned with ethics and social policy since its inception as a discipline. Anthropologists are expected to follow a code of ethics that includes responsibilities such as doing no harm, obtaining informed consent, and maintaining anonymity. And social ethics is a field that addresses issues such as the distribution of economic goods, human rights, animal rights, etc.

Madhuri Sharon. Risk, Safety, and Ethical Concerns in Nanotechnology, (257–276) © 2025 Scrivener Publishing LLC

Some reasons why anthropologists are hesitant to have engaged with ethical issues such as human rights and global inequality include: fear of being similar to universalizing ideologies; and the belief that they were social scientists who could not make moral judgments about other societies. Still, many anthropologists were moved to enter the ethics discipline because of a strong concern for the peoples of the world. During their fieldwork, most have developed a strong empathy for the peoples they have studied and have felt a sense of personal responsibility for their welfare. A primary ethical obligation shared by anthropologists is to do no harm.

The serious concern about ethics and anthropology is followed by different civilizations; and contemporary concerns have been reviewed by many from different angles [1–10].

More recently, a useful survey of ethics in anthropology was contributed by Fluehr-Lobban [11, 12]. Hence, I don't feel the need to repeat these historical efforts that raised the importance of social ethics and anthropological concerns for evolving science, technology and its applications.

The nanotechnology which has opened the door of possibilities to technical wonders has also opened the door to many ethical issues regarding the responsible development of nanotechnology. In this chapter, I will try to look into the social ethics issues and anthropological concerns related to how the entrance of nanotechnology in our daily lives and environment is affecting us. Nanoethics addresses the societal and ethical implications (SEI) of the advancement of nanotechnology/science as it impacts humans, society and the environment.

12.2 Social Ethics and Nanotechnology

Social ethics is the study of moral principles and values that govern human behavior within society. It examines how individuals and groups should behave in relation to each other, considering what is right and wrong, just and unjust, and fair and unfair in social contexts. It includes topics like justice, equality, human rights, responsibility, and the common good. Social ethics is also concerned with issues such as poverty, racism, gender equality, and environmental justice, among others. Social issues that it is involved with are problems that affect a large number of individuals within a society. They often involve societal structures and institutions and can have wide-ranging impacts on the community or nation. For example, common social issues include inequality, unemployment, crime, discrimination, education disparities, and public health challenges [16].

The application of nanotechnology has significant implications for social ethics, touching on areas like privacy, equity, health, environmental impact, and regulatory frameworks. Following is a breakdown of some key ethical concerns.

12.2.1 Health Safety Risks and Social Ethics Due to Nanotoxicity

Due to their size, nanoparticles can interact with biological systems in unpredictable ways, potentially leading to toxicity. Ethical concerns arise when products incorporating nanotechnology are released without full understanding of their long-term health effects on humans. Consumers might not be fully aware of nanomaterials in products, raising concerns about transparency and their right to know the risks associated with the technology. Several health and safety risks due to nanomaterials include toxicity to human tissues and cells, which can cause cell death, DNA mutation, and structural damage to mitochondria. Organ damage is another possibility, including permanent cell damage to the brain, heart, liver, kidneys, spleen, bone marrow, and nervous system. Some studies suggest they can lead to lung cancer due to inhaled nanoparticles that are harder to rinse out than larger particles, which can lead to lung damage [13]. Nanomaterials can unexpectedly become chemical catalysts and cause unanticipated reactions, such as fires and explosions.

Some ethical issues related to nanotechnology include identifying and communicating risks. The process of identifying hazards is the basis for risk management decisions. Workers must consider and accept the certainty and severity of the risk, the health effects, and the compensation they may receive. There is an ethical question of whether enough controls are being put in place to prevent injury and illness. Here, medical screening can help detect disease in its early stages. This will require research investment in toxicological and control research. There is a debate about tolerating the potential for harm to some in the context of anticipated benefits to society.

12.2.2 Environmental Impact of Nanomaterial Causing Ecological Risks and Concern for Social Ethics

Nanomaterials may accumulate in the environment, affecting ecosystems. Nanotechnology may lead to unintended consequences like pollution, which raises ethical questions about precautionary principles and responsibility. Nanomaterials reaching into the land have the potential to contaminate soil, and migrate into surface and ground waters. Particles in solid wastes, wastewater effluents, direct discharges, or accidental spillages can be transported to aquatic systems by wind or rainwater runoff. Biomarkers such as soil samples and zebrafish larvae are used to investigate the environmental effect of nanoparticles. This work has revealed the toxic effects of nanoparticles to include production of oxidative stress, DNA damage, apoptosis, cell death, and induction of inflammatory responses [14]. The potential for nanotechnology to improve energy efficiency and reduce waste could benefit sustainability efforts, but ethical issues arise if the focus is only on profits rather than long-term environmental stewardship.

12.2.3 Privacy and Nano-Surveillance Related to Social Ethics

Nanotechnology's potential for undetectable surveillance and other ethical issues raises concerns about privacy and social ethics. Some nanotechnology is developed for surveillance purposes, which can pose privacy risks, such as in the workplace. Nanosensors used in the workplace can raise ethical concerns about employee privacy, security, and autonomy. Nanotechnology research has focused on ethical issues like privacy, security, equity, environmental impact, and human-machine interactions. Some believe that ethical reflections should be part of the research process from the beginning, rather than being an afterthought. Advanced nanosensors can be embedded in everyday objects, potentially allowing for widespread, unnoticed surveillance. This raises ethical concerns about privacy and the potential misuse of data collected through nanotechnology. Individuals might not be aware of nanotechnology's role in tracking personal behavior or health, challenging ethical standards around informed consent and the right to control personal data. Above all, comes stakeholder interests. When epidemic surveillance goes digital, new stakeholders emerge, such as tech companies and telecommunication companies. These stakeholders may have different ethical standards for data management and use than national governments and health agencies.

12.2.4 Social Inequality: Access to Benefits of Nanotechnology Related to Social Ethics

Nanotechnology could deepen global inequalities if the benefits (e.g., healthcare, cleaner energy, or materials) are only accessible to wealthy countries or corporations. Ethical concerns focus on ensuring that nanotechnology's benefits are distributed fairly. In developing countries, access to nanotechnology may be limited, exacerbating existing disparities. Ethical issues also arise in the prioritization of research—whether it's geared toward solving pressing global problems (like clean water) or enhancing luxury goods.

According to Kamarulzaman *et al.* [15], public benefit and risk perceptions of nanotechnology development are related to psychological and sociological aspects. Today's public is worried about uncertain risks associated with nanoproducts. One has to understand public perceptions of nanotechnology development for effective policymaking. Public culture, religious beliefs and social groups influence benefit perception but not risk perception. However, due to product marketing the public has started seeing nanotechnology as more beneficial than risky, thus increasing benefit perception.

12.2.5 Economic and Employment Disruption or Job Displacement Due to Monopolization of Knowledge About Nanotechnology

As nanotechnology advances, it may lead to automation and significant changes in manufacturing, potentially displacing workers. Ethical considerations include how society prepares for and supports workers affected by these technological shifts. Large corporations may patent nanotechnology breakthroughs, limiting competition and innovation, which can lead to monopolistic control over important technologies.

Nanotechnology isn't just about tiny particles and lab coats; it's a powerhouse of economic potential. Let us look at its impact on economic growth. Nanotechnology advances, fueled by the National Nanotechnology Initiative (NNI), are changing our society. In the US, the NNI has contributed at least $42 billion to the economy in a single year! Small science, big economic impact! Nanotech is not just about robots and quantum dots; it also is about jobs. Nanotechnology firms contribute to employment growth. Regions rich in knowledge become incubators for these firms, stimulating job opportunities. But specialization can be a double-edged sword. Both big and older firms and small younger ones are contributing toward its growth. The gap between those who control nanotechnologies and those whose jobs might be displaced by them is a tug-of-war situation. Will nanotech exacerbate existing inequalities? Or will it democratize progress? The question continues.

Ethical considerations are forcing us to ponder risk awareness, because risks lurk beyond lab mishaps to societal health, privacy, and unintended consequences. How do we balance innovation with societal well-being? Can we enjoy breakthroughs while respecting ethical boundaries?

Nanotech shifts power dynamics between citizens vs. governments vs. corporations. Who leads? Who follows? Who changes the rules?

So, as we pirouette through the nanosphere, let's keep our eyes on the prize: responsible development. Transparency, public trust, and thoughtful policies lead to ethics, which matters.

Despite many benefits of nanotechnology there are potential risks and ethical issues involved in its implementation.

12.2.6 Nano-Related Regulation and Governance Encompassing Regulatory Oversight

Nanotechnology's rapid development has outpaced regulatory frameworks in many cases. Ethical issues arise from the need to ensure that adequate regulations are in place to protect public health, safety, and the environment without stifling innovation. Discrepancies in how different countries regulate nanotechnology create ethical challenges, especially regarding international trade and the potential for dumping of harmful products in countries with weaker regulatory systems.

Regulation and governance of nanotechnology vary across different countries and regions, making the establishment of a global standard challenging. These differences arise due to varying levels of technological advancement, public perception of nanotechnology, legal frameworks, and national priorities. Christopher Bosso [17] proposed settling into the midstream. That is, lessons for governance from the decade of nanotechnology.

A cursory glance at how regulatory oversight and governance operate and how they differ globally:

12.2.6.1 Regulatory Oversight at the National and Regional Level

In the US, nanotechnology is regulated under existing frameworks, such as the Toxic Substances Control Act (TSCA) and the Food, Drug, and Cosmetic Act (FDCA), through agencies like the Environmental Protection Agency (EPA), Food and Drug Administration (FDA), and Occupational Safety and Health Administration (OSHA). The National Nanotechnology Initiative (NNI) is a federal program focused on coordinating nanotechnology R&D. Whereas the European Union has a precautionary approach to nanotechnology. It incorporates nanomaterials into existing regulations such as REACH (registration, evaluation, authorisation, and restriction of chemicals) and cosmetic products Regulation (EC) No 1223/2009. It also mandates the labeling of nano-based products in food and cosmetics.

Japan emphasizes voluntary industry standards, supported by regulatory frameworks like the Chemical Substances Control Law. The National Institute for Materials Science (NIMS) and Ministry of Economy, Trade, and Industry (METI) also play roles in nanotech regulation.

China is rapidly advancing its nanotechnology sector and has been integrating nanotechnology into its existing regulations for chemicals and consumer products. There is significant involvement from the government, though specific nano regulations are still developing.

In India, regulatory oversight for nanotechnology is still developing and operates under existing frameworks for chemicals, consumer goods, environmental protection, and health and safety. While India has made significant strides in research and development within nanotechnology, dedicated and comprehensive regulation is still emerging. Here, at the national level the Ministry of Science and Technology has launched the Nano Mission in 2007, which is the principal program for advancing nanotechnology in India. The Nano Mission focuses on: infrastructure development for R&D; human resource development; international collaborations; and public-private partnerships in nanotechnology development. However, the Nano Mission is more research-oriented and does not directly address the regulatory aspect of nanotechnology. Other departments involved are the Ministry of Environment, Forest, and Climate Change (MoEFCC). The Environment Protection Act (1986) is a broad piece of legislation that covers the regulation of activities, including those involving nanomaterials, that may impact the environment. The Central Pollution Control Board (CPCB) plays a role in regulating the environmental and health safety of industries that might manufacture or handle nanomaterials. The Ministry of Chemicals and Fertilizers; Central Drugs Standard Control Organization (CDSCO), Food Safety and Standards Authority of India (FSSAI), and Bureau of Indian Standards (BIS) work in collaboration with international bodies like the International Organization for Standardization (ISO) to develop these standards. Moreover, institutions like the Indian Institutes of Technology (IITs), Council of Scientific and Industrial Research (CSIR), and other academic bodies are actively researching the potential health, environmental, and safety risks posed by nanotechnology.

12.2.6.2 Challenges in Global Standardization

There is a divergence in approaches by different countries. Countries adopt either a precautionary approach (like the EU) or a more flexible one (like the US), which complicates global harmonization of standards.

Because there is no universally accepted definition of what constitutes a "nanomaterial," which leads to inconsistency in regulations, international collaboration initiatives, such as the OECD Working Party on Manufactured Nanomaterials (WPMN) and the International Organization for Standardization (ISO), are working towards harmonizing nanotechnology standards globally. However, the adoption of such standards is not uniform across all countries. The rapid pace of innovation in nanotechnology means that regulations may lag behind scientific advances, leading to gaps in oversight. Different countries have varying levels of concern regarding the potential risks posed by nanomaterials to health and the environment. This can result in stricter regulations in some regions compared to others.

12.2.6.3 Potential Pathways for Global Governance

International Standards and Guidelines organizations like the ISO and OECD are crucial for developing common guidelines for testing, risk assessment, and labeling of nanomaterials. Their guidelines could eventually lead to more cohesive global standards. For transparency and data sharing, increased collaboration between countries in terms of research data, safety evaluations, and regulatory practices could help streamline oversight processes.

As nanotechnology develops, global governance frameworks may need to address broader ethical concerns, including environmental justice, privacy, and equity in access to nano-based innovations.

The regulation of nanotechnology will continue to evolve, and international coordination will be key to ensuring the safe and equitable development of this transformative field.

12.2.7 Nanotechnology Led Human Enhancement: Ethical Dilemmas in Human Augmentation

No doubt, where nanotechnology meets human enhancement is a captivating realm in which intricate ethical dilemmas arise—especially when it comes to identity and personhood.

Nanotechnology has potential applications in enhancing human abilities, from cognitive improvements to physical augmentation. Ethical concerns center around the fairness of these enhancements, consent, and the potential for creating a societal divide between "enhanced" and "non-enhanced" individuals. Alterations to the human body through nanotechnology challenge concepts of identity, autonomy, and what it means to be human, prompting deep ethical reflections. Let us enquire into nanotechnology and human enhancement; and ethical implications for identity and personhood.

12.2.7.1 Nanotechnology and Human Enhancement

Nanotechnology, with its promise of manipulating matter at the atomic and molecular scale, has opened doors to enhancing human capabilities. But as we tiptoe toward a future where we can modify ourselves at the nanoscale, we find ourselves engrossed in many concerns acting together. Nanotechnology has convergence of nanotechnology, biotechnology, information technology, and cognitive science (NBIC) functioning together. This convergence provides the basis for manipulating nearly every aspect of human functioning. From nano-enhanced red blood cells to neural implants, we are on the cusp of reshaping what it means to be human.

Traditionally, medicine has focused on treating existing diseases or injuries. But emerging technologies blur the line between therapy and enhancement. Is there a practical or ethical difference between injecting nano-enhanced cells and providing oxygen? Or between human organ transplants and nano-engineered organs? We are treading on the edge of a transformative activity.

Also, what autonomy and choice do we have? Should individuals have the autonomy to pursue enhancement? Will compulsory technologies limit our choices? And what about those who choose not to enhance—will they become a vulnerable population? These questions echo through the gathering of experts.

12.2.7.2 Ethical Implications for Identity and Personhood

Our personal identity is in flux. This is the heart of the matter—the very essence of who we are! When we enhance ourselves, do we alter our core identity? Imagine nano-enhanced memory implants—preserving memories but potentially reshaping our sense of self. As we dance ahead with enhancements, we must grapple with questions about continuity, authenticity, and what remains uniquely "us." Then the question of the extended self comes up. If we extend our capabilities—whether through nanotech or other means—does our sense of self expand too? Are we still the same person when we can compute faster, remember more, or even live longer? The dance floor trembles with uncertainty. As some embrace enhancements, others may resist. Society swirls with debates: Should we prioritize collective well-being or individual autonomy? Will enhanced humans and non-enhanced humans coexist harmoniously, or will divisions deepen? This creates social and ethical ripples. So, we must tread carefully. Our steps ripple through time, affecting generations to come. So, as we pirouette through nanotech and human enhancement, let's keep our hearts open to both wonder and responsibility.

12.2.8 Nano-Weaponization and Dual-Use Dilemmas in Military and Security: An Ethical Concern

Nanotechnology is being used in the development of new military technologies, such as advanced weapons or surveillance tools. This raises ethical concerns about the responsible use of nanotechnology in conflict situations and the potential for escalation in arms races. Many nanotechnologies have both civilian and military applications, leading to ethical concerns about their potential misuse for harm, even if initially developed for peaceful purposes. Though, nanotechnology offers tremendous potential benefits, it also raises complex ethical issues. Ensuring that nanotechnology is developed and applied in ways that promote social good, protects individuals and the environment, and maintains fairness is a crucial ethical challenge for researchers, policymakers, and society at large.

12.2.8.1 Nano-Weaponization

The concept of nano-weaponization refers to the use of nanotechnology in the development of military or security applications, such as weapons and defense systems. Nanotechnology, which operates at the molecular or atomic level, has vast potential for innovation but also raises significant ethical concerns, especially when used for dual-use purposes—applications that can be both beneficial for civilians and dangerous in military contexts.

12.2.8.2 Dual-Use Dilemma

Dual-use technologies are those that can be applied to both civilian and military purposes. In the context of nanotechnology, there are several dilemmas that arise from use of nanotechnology for the following purposes:

(i) *Medical and Military:* Nanotechnology can be used to enhance drug delivery systems, but similar systems could also be weaponized to deliver toxins or viruses in warfare. Research intended for disease treatment could be repurposed for biological warfare.

(ii) *Surveillance vs. Privacy*: Nano-sensors could be developed to monitor environmental pollutants or detect chemical weapons, but these sensors could also be used for mass surveillance, infringing on privacy rights.

(iii) *Defense vs. Offense*: Nanomaterials can improve the strength and resilience of military gear or vehicles, which could aid in defense. However, they can also be used to create more powerful and stealthy offensive weapons, potentially leading to an arms race.

12.2.8.3 Ethical Concerns

Ethical concerns due to use of nanotechnology in the military may cause:

(i) *Escalation of Conflict*: The development of nanoweapons could make conflicts more destructive, with the potential for mass casualties. This raises concerns about the **just** war theory, particularly the principles of proportionality and distinction, which are fundamental in ensuring that military action is ethical.
(ii) *Unpredictable Long-Term Effects*: Nanotechnology operates at scales that could have unpredictable biological, environmental, and societal impacts. Ethical questions arise regarding the unintended consequences of these technologies, especially if their use results in harm to civilians or the environment.
(iii) *Lack of Regulation*: There is currently insufficient global regulation or treaties specifically addressing nanotechnology in warfare. This could lead to an unchecked arms race or misuse of technology by rogue actors. International governance structures need to catch up to address these concerns.
(iv) *Accountability*: The dispersal of nano-sized weapons or devices could make it difficult to trace responsibility. If these weapons are used by non-state actors or in covert operations, it could be difficult to hold the perpetrators accountable, further complicating ethical evaluations of conflict.
(v) *Human Enhancement:* Nanotechnology could also be used for the enhancement of soldiers, potentially leading to a future where enhanced humans create new ethical and legal challenges, particularly around human rights and equality.

12.2.8.4 Possible Solutions and Ethical Guidelines

The ethical concern over nano-weaponization and dual-use dilemmas is highly relevant in today's technological and geopolitical landscape, where the line between civilian and military applications continues to blur. The challenge lies in harnessing nanotechnology for the benefit of humanity while preventing its misuse in harmful ways. Hence, there is a need to have

(i) *International Treaties and Regulation*: Just as chemical and biological weapons are regulated by international treaties, there needs to be a concerted effort to regulate nanotechnology in the military and security domains.
(ii) *Transparency and Oversight:* More transparent oversight mechanisms, perhaps through international organizations, could help ensure that nanotechnology is not misused.
(iii) *Ethical Frameworks:* A global ethical framework should be established to govern the use of nanotechnology, focusing on human rights, peace, and sustainability.
(iv) *Dual-Use Research Governance:* Research in nanotechnology should have strict protocols in place to prevent civilian applications from being easily converted into military uses.

12.3 Anthropological Questions Pertaining to Nanotechnology

Anthropological questions pertain to the study of human beings, their behaviors, cultures, and societies across time. Anthropology is the discipline that seeks to understand the complexities of human existence by examining cultural, social, biological, and linguistic aspects of humans.

Some anthropological questions include:

- How do different cultures interpret the world?
- What are the origins of human societies?
- How do social norms and values evolve?

The intersection of nanotechnology and anthropology is a fascinating aspect to delve into in this intriguing realm where the infinitesimal meets the cultural. Nanotechnology, with its promise of manipulating matter at the atomic and molecular scale, has indeed sparked questions about what it means to be human, our relationship with materials, and the boundaries between nature, technology, and ourselves. Here are some anthropological questions that arise in the context of nanotechnology:

(i) *Posthumanism and Immortality:* Is it in our nature to become posthumans? As we explore nanoscale manipulation, we inevitably ponder our own transformation. Will we enhance ourselves to defy illness, aging, and even mortality? The immortality society awaits us—how will we navigate it?
(ii) *Material Agency and Cognition*: Nanotechnology blurs the lines between what's intrinsically natural, technological, or human. How do we perceive agency in materials that can self-assemble, respond to stimuli, or even compute? Can we anthropomorphize these tiny building blocks?
(iii) *Visionary Elements and Cultural Meanings*: Just as medieval alchemists sought the ultimate secret of mastering matter and spirit; visionary nanotechnologists aim for "molecular manufacturing." But what cultural meanings do we attach to this quest? How does it shape our understanding of ourselves and our place in the cosmos?
(iv) *Human-Machine Transmutations*: Nanotechnology isn't just about creating intelligent artifacts; it's also about enhancing humans. How do we navigate the boundary between our biological selves and the engineered enhancements we might embrace? Are we becoming cyborgs, blending the organic and the synthetic?
(v) *Ethical and Social Implications:* As nanotechnology advances, we must grapple with ethical dilemmas. How do we ensure equitable access to these transformative technologies? What cultural norms and values guide our decisions about their deployment?

12.4 Uncertainties in Detection (ID Detection) Concerning Nanotechnology

This term seems to refer to uncertainties or challenges in detecting or identifying something, possibly in a technological or analytical context. "ID detection" could relate to identity detection, which involves verifying the identity of individuals, often through technology like biometrics, IDs, or digital verification systems. In broader terms, this might involve uncertainties in verifying identities due to issues like data accuracy, privacy concerns, or technological limitations.

Uncertainties in detection, particularly ID detection (identity detection), in the context of nanotechnology refer to the challenges and ambiguities in accurately identifying, measuring, and monitoring nanomaterials, as well as their behaviors and impacts in biological and environmental systems. These uncertainties have far-reaching implications for safety, ethics, and regulation. Some key areas of concern are:

(i) *Detection of Nanomaterials in the Environment:* Nanoparticles come in various sizes, shapes, and compositions, making it difficult to standardize detection methods. Current technologies may struggle to detect nanoparticles consistently due to their small size and variability in form. Besides this, nanoparticles can agglomerate or aggregate, altering their behavior and making them harder to detect individually. This raises uncertainty about their actual concentrations in the environment or biological systems, complicating risk assessments. Moreover, nanomaterials often exist at low concentrations, making them difficult to detect in complex environments like air, water, or soil. Standard environmental monitoring techniques may not be sensitive enough to measure nanoparticles accurately, creating uncertainty in understanding their distribution and effects.

(ii) *Biological Detection and Interaction:* The possibility of complex interactions of nanoparticles with biological systems in unpredictable ways, binding with proteins or cells and changing their detectable characteristics; makes it challenging to assess how these materials behave in living organisms, where they accumulate, or what effects they might have. Nanoparticles might accumulate in tissues or biomagnify through the food chain, but detecting these processes is complex due to the difficulty in tracing nanoparticles across different biological scales and systems. There is uncertainty in how nanoparticles can be reliably detected within human tissues, especially over long periods. Nanoparticles can pass through biological barriers like the blood-brain barrier, but their exact pathways and effects are not well understood, making detection and monitoring challenging.

(iii) *Limits of Current Detection Technologies:* Many existing detection methods, such as spectroscopy or electron microscopy, are not yet refined enough to consistently identify the presence of nanoparticles at low concentrations or in mixtures with other materials. This technological gap contributes to uncertainty in how nanomaterials are monitored. Nanoparticles can interfere with existing detection methods, leading to false positives or negatives. For instance, they might produce signals similar to those of other materials, making it difficult to accurately identify and quantify them. Achieving real-time monitoring of nanomaterials remains a major challenge. The lack of continuous, real-time detection methods increases uncertainty in assessing exposure levels and immediate impacts.

(iv) *Regulatory and Standardization Challenges:* There is no universal agreement on the best methods for detecting and measuring nanoparticles, which creates inconsistencies across research studies, industries, and regulatory frameworks. This lack of standardization results in significant uncertainty regarding safety thresholds and acceptable exposure levels. There is an ongoing debate about how to best measure exposure to nanomaterials. Should exposure be based on the number of particles, their total surface area, or their mass? This uncertainty in metrics complicates detection and risk assessment. Moreover, to distinguish between engineered nanoparticles (ENPs) and naturally occurring nanoparticles in the environment can be difficult. Without clear detection methods, there's uncertainty about which materials are being assessed and which may pose risks.

(v) *Detection of Toxicity and Long-Term Effects:* Nanoparticles may cause subtle or cumulative biological effects that are hard to detect over short-term studies. This uncertainty makes it difficult to assess the long-term health or environmental risks of exposure to nanomaterials. Plus, it is unclear how nanoparticles induce toxicity in many cases. Current toxicological tests might not capture the full range of effects, leading to incomplete detection of potential hazards. This raises ethical concerns about releasing products containing nanomaterials without fully understanding their risks.

(vi) *Privacy and Ethical Concerns in Nanoscale ID Detection:* Nanotechnology enables the creation of highly sensitive nanosensors that can detect a wide range of biological and environmental signals. This could include the detection of personal biological markers, raising privacy concerns about how this information is collected, stored, and used, especially without individuals' consent. When nanotechnology is used for identification or surveillance purposes (e.g., in healthcare or security), it may become difficult for individuals to give informed consent, as they may not be aware of the presence or capabilities of nanoscale detectors. Nanosensors embedded in devices for tracking health, behavior, or location could compromise personal privacy. Detecting and managing data from these sensors creates uncertainties around data security, the risk of misuse, and ensures anonymity.

(vii) *Detection in Regulatory and Ethical Frameworks:* Uncertainty in the detection of nanomaterials complicates efforts to enforce regulatory standards. Without reliable detection, it's challenging to ensure compliance with safety regulations, leading to ethical dilemmas regarding public and environmental health. Scientists and manufacturers face ethical dilemmas when releasing nanotechnology-based products into the market without fully understanding or being able to detect their long-term impacts on health and the environment. This uncertainty raises questions about the precautionary principle and who bears responsibility for potential harm.

The uncertainties in detection surrounding nanotechnology pose significant ethical, safety, and regulatory challenges. Current detection technologies are often inadequate for consistently identifying nanomaterials in complex environments, accurately assessing exposure, and predicting long-term effects. These challenges emphasize the need for improved detection methods, standardized protocols, and ethical considerations in the development and application of nanotechnology to ensure responsible use and minimize risks.

12.5 Conclusion

Looking at the ethical concerns that we have discussed in this chapter, it is better to ask about the advances in nanotechnology: *Should we do this*? rather than *Can we do this*?

The lack of clarity about the hazards of nanomaterials and the risks requires an in-depth assessment of the hazards and risks that might exist in various situations, so that it can be viewed through the window of social ethics and anthropological concerns. The question arises as to what is being done and what is missing? As questioned by Lu *et al.* [18]. Both employers and workers should be able to exercise their autonomy only if the processes leading to hazard identification and risk assessment are transparent and understandable and conform to the principles of autonomy, beneficence, nonmaleficence, justice, privacy and respect for persons. This will enable them to accurately portray hazards and risks, offer a precautionary approach to hazards, and take the necessary steps to control risks so that they appear reasonable and acceptable to workers/researchers.

Considerable variability in the types of nanomaterials, their chemical composition, structure, and functionalization, poses relatively few options for exposure control and determining in which hazard and control "bin" a nanomaterial fits. Additional information may be obtained through comparative toxicity to benchmark (reference or control) particles. Such a strategy may also facilitate comparison across particle types and development of safer nanomaterials. A key challenge, in order for nanotechnology to deliver on its promise of societal benefit, is to implement effective measures to reduce or eliminate occupational exposures is an early step in a responsible life-cycle approach and the primary approach needed to prevent adverse health effects in workers producing or using nanomaterials. In conclusion, it can be said that anthropology plays a crucial role in deciding social ethical concerns by providing a deep understanding of different cultures and societies, allowing researchers to analyze practices and beliefs from a nuanced perspective, which can then be used to identify potential ethical issues and inform responsible action when encountering diverse social norms and values across different communities; which often involves considering concepts like cultural relativism while still maintaining a commitment to "do no harm" in research practices.

References

1. Boas, F., Correspondence: Scientists as spies. *Nation*, 109, 2842, 1919.
2. Alexander, MB,., *Experiments in Living: A Study of the Nature and Foundation of Ethics or Morals in the Light of Recent Work in Social Anthropology*, Macmillan, London, 1952.
3. Brandt, R.B., *Hopi Ethics: A Theoretical Analysis,* University of Chicago Press, Chicago, 1954.
4. Edel, A., *Ethical Judgement: The Use of Science in Ethics*, Free Press, Glencoe, IL, 1955.
5. Ladd, J.T., *The Structure of a Moral Code: A Philosophical Analysis of Ethical Discourse Applied to the Ethics of the Navaho Indians,* Harvard University Press, Cambridge, 1957.
6. Tylor, E.B., *Primitive Culture. Volume 11: Religion in Primitive Culture*, Harper Torchbook, New York, 1958, 1871.
7. Srinivas, M.N. (Ed.), *Method in Social Anthropology: Selected Essays by A. R. Radcliffe-Brown*, University of Chicago Press, Chicago, 1958.
8. Edel, M. and Edel, A., *Anthropology and Ethics*, Charles C Thomas, Springfield, IL, 1959.
9. Bidney, D., The concept of value in modern anthropology, in: *Anthropology Today: Selections*, S. Tax (Ed.), pp. 436–453, University of Chicago Press, Chicago, 1962.
10. Maclntyre, A., *A Short History of Ethics. New York: Macmillan.1981 After Virtue: A Study in Moral Theory*, University of Notre Dame Press, Notre Dame, 1966.
11. Fluehr-Lobban, C., Ethics, in: *Handbook of Methods in Cultural Anthropology*, H. Russell Bernard (Ed.), pp. 173–202, Walnut Creek, CA: AltaMira, 1998.
12. Fluehr-Lobban, C., A century of ethics and professional anthropology. *AAA Anthropol. News*, 43.3, 20, 2002.
13. Lihui, X., Ju, Z., Skonieczna, M., Zhou, P.-K., Huang, R., Nanoparticles-induced potential toxicity on human health: Applications, toxicity mechanisms, and evaluation models. *MedComm (2020)*, 4, 4, e327, 2023 Aug. Published online 2023 Jul 14. doi: 10.1002/mco2.327. PMCID: PMC10349198PMID: 37457660.
14. Guillermo, M., Merinero, M., Pérez-Aranda, M., Pérez-Soriano, E.M., Ortiz, T., Begines, B., Alcudi, A., Environmental impact of nanoparticles' application as an emerging technology: A review. *Mater. (Basel)*, 14, 1, 166, 2021 Jan. Published online 2020 Dec 31. doi: 10.3390/ma14010166. PMCID: PMC7795427PMID: 33396469.
15. Kamarulzaman, N.A., Lee, K.E., Siow, K.S., Mokhtar, M., Public benefit and risk perceptions of nanotechnology development: Psychological and sociological aspects. *Technol. Soc.*, 62, 101329, 2020, ISSN 0160- 791X, https://doi.org/10.1016/j.techsoc.2020.101329.

16. Khan, A., Ethical and social implications of nanotechnology. *QScience Proceedings (Engineering Leaders Conference 2014)*, vol. 57, 2015, http://dx.doi.org/10.5339/qproc.2015.elc2014.57.
17. Bosso, C., Settling into the midstream? Lessons for governance from the decade of nanotechnology. *J. Nanopart. Res.*, 18, 163, 2016, DOI:10.1007/s11051-016- 3451-9.
18. L.Y.Y., Lu, Lin, B.J.Y., Liu, J.S., Yu, C.-Y., Ethics in nanotechnology: What's being done? What's missing? *J. Bus. Ethics*, 109, 4, 583–598, 2012.

Epilogue

Science, Technology, and Ethics: The Modern Interface

In this book, an effort has been made to understand the ethical quandaries surrounding nanotechnology through a general scientific perspective as well as its more precise and unique perspective.

Science, Technology, Social Science and Human Values: Integrated Perspectives Are of Concern

The credence of today is that scientific and technological progress can have a profound impact on every aspect of life in a society. From the very existence of atom bombs dictating the larger geo-political policies of global powers to the artificial intelligence on your mobile devices that categorizes your photos based on the people in it, science influences macro and micro behaviors of the masses, in both the short and long-term. It is then clear that unethical application of science and tech leads to large-scale disasters, and this very fact helps society decide and enforce morality on progress.

For example, in the last fifty years global warming has become a physical threat to humanity, and economic disparity has wreaked havoc on the poor and has become a problem to the prevailing regimes. When this occurs, we are facing economic crises like recession, unemployment, and price hikes; and bewilderment; and psychological crises like confusion, anxiety, depression, and crimes. It is increasing dissatisfaction among the "voters" and "consumers" at a larger scale. This has led many businesses, which were operated in an unethical way, to collapse due to public outcry. In addition to that, public outcries have helped guide policy at government level, such as when world governments collectively banned adding lead to petrol fuels once the horrible birth defects were linked to the lead-compound pollution. Consequently, some sort of economic help, philanthropy and corporate social responsibility (CSR) concepts have been developed to pacify the political and economic problems. But often, these concepts and their executions are self-contradictory and unrealistic and effectively a farce, which do not deliver any sustainable solution to such an ever-growing set of ethical problems. Examples for these actions abound, such as a company advertising false metrics to claim "green production" by not accounting for the carbon footprint from their overseas operations.

Thus, it is crucial for us to understand how a science as powerful as nanoscience can and should be controlled. To do so, let us understand what nanoscience and nanotechnology can do, and what makes it a uniquely challenging science to monitor and control.

Wang *et al.* [1] has asserted that the ancient Chinese philosophy of "unity of man with nature" provides principles which can guide and develop human values into a new, positive force with the potential to harmoniously manage sustainable ecosystem services.

Nanoscience and Nanotechnology: A New Revolution

Nanoscience and nanotechnology represent a new scientific and technological revolution. Nanotechnology will alter our relationship with molecules and matter as profoundly as the computer changed our relationship with bits and information [2]. It is molecular manufacturing; it operates at a size range less than 100 nanometers or billionths of a meter. It transforms from huge to nano, from visible to invisible, from limited space to unlimited space, from slow speed to fast speed, from expensive to affordable, from individual consumption to large community consumption. In nanotechnology, we can fabricate machines and devices at the atomic level. With this ability, we would be able to produce molecular-sized computers and robots with unprecedented control over matter as we see desirable.

The Old Legacy of Nanotechnology and Today's Perspective

Existence of nanomaterial from prehistoric to contemporary period has been well documented [3]. While nanotechnology is regarded as one of the most cutting-edge fields today, it has a legacy as old as nature itself. The structure of organelles of the cells and their functioning in the living body is the most complicated example of functioning as a living organ at such a nanoscale. Neurons and nervous systems are highly sophisticated nanonetworks in our body, which make use of protein interactions to perform learning behavior for an organism. The abalone, and many other marine snails, create the shell using calcium carbonate from the saline sea water by mixing carbohydrates and protein and forming a nanostructured tile, where pearl-like smooth surfaces are created. Dragline silk is also created and weaved by a spider, which is a miracle of strength and ductility itself, boasting higher strength than highest-grade steel alloys. We have also been producing Kajal (black oily powder produced by burning the oil of mustard, coconut, etc., and used for eyes as cosmetics), due to its nanosize smoothness and effectiveness. Kajal has been very popular in South

Asia, the Middle East, and North Africa since 3100 BC. Indian ayurvedic medicines like gold ash (Swarn Bhasm), silver ash (Rajat Bhasm), copper ash (Tamra Bhasm), pearl ash (Mukta Bhasm), iron ash (Lauha Bhasm), and shell ash (Shankh Bhasm) have been used for treating a wide spectrum of bacterial infections (Figure 1). All these ashes are prepared by burning the materials at high temperature, through a prescribed process that controls the supply of oxygen, transforming into a colloidal form, ~8-10 nm, which is on the order of the smallest modern nanoparticle.

The effects of these medicines are well established in the Ayurveda method (an Indian old method of treatment of diseases), being used for the last thirteen hundred years and is still being used effectively. Today, we can prepare semiconductor, plasmonic, and magnetic nanomaterials using the various emerging nanotechnological processes, like synthesis of nanomaterials through physical, chemical, and biological methods, self-assembly, analyzing the creations by microscopic techniques and diffraction techniques. We can have special nanomaterials like carbon, porous, aerogel, zeolites, core-shell particles, metamaterials, bioinspired materials, etc. With these technologies, we can even hope to boost energy production and distribution, through dye-sensitized photovoltaic solar cells, organic photovoltaic cells, fuel cells, hydrogen production and storage technology, and hybrid energy cells. The materials used in automobiles, sports, toys, textiles, and cosmetics are also being supported by nanotechnology.

Nanotechnology as a Boon to Solve Economic Problems

Unfortunately, about 25% of the world population is still living in very poor conditions. We are facing a food shortage, water shortage and contamination, environmental hazards, garbage heaps on land and sea, unemployment, inflation, scarcity of infrastructure — roads, sewerage, transportation, communication, and maldistribution of energy in many

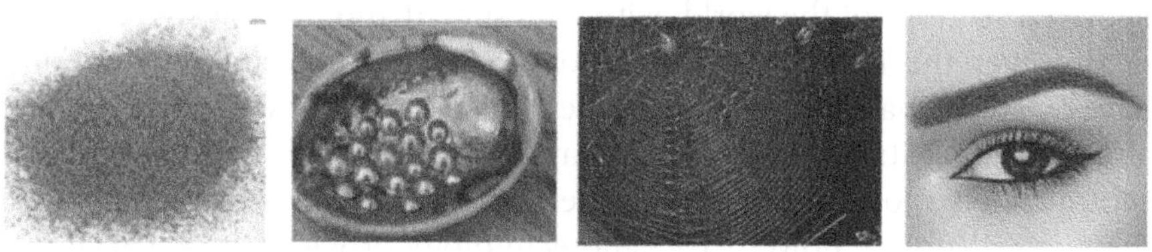

Swarna Bhasma (Gold ash) *Abalone Pearl* *Spider web* *Kaajal (Kohl)*

Figure 1 Man-made nanomaterial (Swarna Bhasma containing gold nanoparticle), natural nanoparticle-containing material (Pearl & Spider web) and Kaajal, an eye cosmetic prepared using carbon nanomaterial.

countries of Africa, Asia, and South America. Thus, nanotechnology has a great role in solving most of these problems. This technology provides fundamentally new types of pharmaceuticals, semiconductor devices, sensors, renewable energy, security capture and storage systems, water purification devices, super strong fabrics and materials, security, and military components, as well as anti-pollution devices. These applications are already emerging and will strengthen over the years [4].

In the medical field, imaging, drug delivery, cancer therapy, and tissue repair are being supplemented by nanotechnology. It is transforming agriculture and food, domestic appliances, space, defense, and engineering. This is nanotechnology replacing the "Big," and the "Huge" by "Nano" at the micro level, widely and deeply [5].

At the end of the first quarter of the first century, we have to respond to problems due to increased or decreased population in different countries of the world, such as deflection of water, fossil oil, coal-like natural resources and other environmental problems; thus, replace and supplement them by technologically advanced and economically cheaper resources. If nanotechnology is the answer to these huge problems, then the best wisdom of humanity is to be dedicated to this useful scientific technology. Nanotechnology is the richest plateau of the scientific revolution, where all humanity has to reside at the moment at our doorstep.

Nanotechnology may appear as a boon considering the scale and speed of nanoproduction and distribution in the world economy. If we can produce energy at cheaper levels and in abundance, many economic facilities will be available at lower prices. In the past, the industrial revolution has provided previously scarce facilities and goods to a larger number of people at increasingly affordable prices. We could provide a fan and light to every poor consumer, which was not available even to kings in medieval times. Now, we need another push of sophisticated technology like nanotechnology, which would serve basic needs of the general masses, especially the deprived sections of the world, with widespread and inexpensive care.

However, this is easier said than done. Science and technology not only have created devastating tools like chemical and atomic weapons, but have also subtly negatively affected the way that power and money have been centralized in most modern economies, despite a democratic setup to deter the same. With the power that nanotechnology provides, it is crucial that we learn from the past to design better policies to mitigate a similar pattern emerging.

However, We Cannot Overlook Potential Dangers of Nanotechnology

While all technologies are double-edged swords, nanotechnology's extraordinary power requires us to think more carefully about the negative consequences. The production of miniature weapons and explosives and disassemblers for military use is possible; rampant nanotechnology can be prepared causing a gray goo scenario in which out-of-control self-replicating processes may end up consuming the entire life on Earth; extensive surveillance could monitor and track the activities of people considered opponents of a state, be it criminals or the political opposition. Accustomed to living with almost routine scientific breakthroughs, we have yet to come to terms with the fact that the most compelling 21st century technologies—robotics, genetic engineering, and nanotechnology—pose a different threat than the technologies that have come before. A bomb is blown up only once— but one bot can become many, and quickly get out of our control. Uncontrolled self-replication in these newer technologies runs a much greater risk: a risk of substantial damage to the physical world. Most dangerously, for the first time, these accidents and abuses are widely within the reach of individuals or small groups. They will not require facilities or rare raw materials. Knowledge alone will enable their use. If we can extend the present capabilities by miniaturizing guns, explosives, and electronic missiles, they will run wild and wreak havoc. Thus, we have the possibility not just of weapons of mass destruction but of knowledge-enabled mass destruction (KMD); this destructiveness is hugely amplified by the power of self-replication. Nanotechnology users would be able to use molecular sized microphones, cameras and homing beacons to explore the secrecy of the others. Thus, nanotechnology may also give us devil-like powers. As Bill Joy says, this is the first moment in the history of our planet when any species, by its own voluntary actions, has become a danger to itself as well as to vast numbers of others [6].

In such an unprecedented situation, a possible solution may be that the scientists and engineers need to adopt a strong code of ethical conduct, resembling the Hippocratic oath, and that we create policies that encourage whistleblowing as necessary, to minimize the often-associated high personal cost. The misuse of nanotechnology will compel us to think of a strong ethical mindset of masses, the direct stakeholders, NGOs, and the governments.

As mentioned by Sharon [7], are we ignoring the considerations for the dark-side of nanotechnology? We have to also take a fresh look at the possibilities of a deliberate abuse of the technology. There is a need to establish a "nanotechnology safety system" that should encompass toxicological research, hazard identification, and risk and exposure assessment as well as risk management.

Social Aspect of Ethics of Nanotechnology Is a Solicitude

Ethics is basically a culmination of a biological process cultured by the social environment through lakhs of years. As man is the most socio-animal being, his every sort of development, even his very existence and survival, his physical and mental health, depends much more on his socialization process. Every individual development, even the issue of modern individual "patents," are each grounded on our long joint social legacy. These so-called "individual" patents are the culmination of knowledge and techniques developed by the participation of our collective forefathers and foremothers through thousands of years. Our conscious, subconscious and unconscious mind is also a composite product of the long chronology of bio-socio life of our long past evolution. The religion, our gods/or no god, culture (way of thinking and behaving) and civilization (way of living); all our religious, cultural, economic, social, and political organizations, are the evolutionary products of long and constant interaction of masses in the past, after great sacrifice, and payment of a heavy price, by our predecessors. It is impossible to locate any individual thought, behavior or innovation or even so-called Nobel Prize "invention," which can be considered purely "individual," and without a series of past, long, cooperative efforts. All us humans, in a form of "liberal" or "free individual" can consider ourselves as an independent, individual, natural, biological unit; with our "own" body and our "own" mind; but every human mind, as such, is a product of shared experiences of lacs of years. Thus, our so-called "liberty" and "individuality" is not an exclusive independent phenomenon; in fact, it is inherently a "social" composite product also, made of an integrated social fiber, evolving in a long communal co-operative environment.

The "other" is an essential, generic, and always coexisting entity, right from birth to the death, for every "self" of this world; so, our every individual decision should be related to the "other," necessarily, and inherently. Consequently, in the field of science and technology, our every invention, creation and production are essentially an "other"-related action and outcome. Thus, the sociology of ethics is standing on two legs of human decisions; one is— any individual thought and action should not do "harm" to the "other"; and second, it should always be in the "welfare" of the "other." Generally, we concentrate more on the first part of the ethics only, and the second concept is not only neglected, rather it is considered irrelevant and impractical also. Generally, in "modern" times it seems impossible to think that someone will do his/her business, any economic activity, for the welfare of the "other." Instead, the nanotechnology has inputs and involvement with all the branches of three powerful sciences, i.e., biotechnology, information technology, and cognitive science. "Other" is considered his/her sole "source" of his/her own personal "benefit," and even exploitation. Nanotechnology will have very broad applications across all fields of science and engineering, so it will be an amplifier of the social effects of other technologies. Nanotechnology itself may possibly raise distinctive ethical and social issues in the future. Thus, it will be important to integrate social and ethical studies into nanotechnology developments from their very beginning. It will help manufacturers and governments to make the right decisions when deploying a new technology, maximizing its benefit for society [8].

We should think seriously that the essence of all the cultures, religions, epics, literature, and the messages of all the gods, prophets, gurus, is the "welfare of humanity." It can be a human's prime duty and the source of their joy and happiness. This is the essential essence of socialization, philosophy, and the soul of ethics at large. But the whole corporate mindset is all against it, and aimed to squeeze the maximum "benefit" and "profit" through any sort of means. Generally, it is believed that even by manipulation of the political governance and the economic system, any "liberated" individual is free to gain the "profit" as much as he/she can but maneuvering to escape from the gallows of the law.

There Has to Be Nanotechnology, Ethics, and Sustainable Development

Here, on this point, the basic issue of ethics arises. We are all social beings that must be ethical not for the "other" only, but for our own sake also. By doing welfare to "others" is not a charity anymore, rather it is an essential prerequisite for our own existence and enjoyment. Batson [10] verifies the theory based on research that "we humans have the capacity to care for others for their sakes (altruism)" and "empathy-induced altruism is within the human repertoire." "Empathy-induced altruism is a far more pervasive and powerful force in human affairs than has been recognized. Failure to appreciate its importance has handicapped attempts to understand why we humans act as we do and wherein our happiness lies. This failure has also handicapped efforts to promote better interpersonal relations and create a more caring, human society."

By being selfish, we can gather more money and facilities, but if we delimit our aim of life—profit-making only—then we are bound to create conflicts with others; which would create tension and sorrow for all, in one way or the other. If we have molded our enjoyment in "giving," "scarifying," ""welfare" of the "other," we are bound to be flourished by joy, this way or that way, here or there, now and then. "Volunteer work improves access to social and psychological resources, which are known to counter negative moods such as depression and anxiety.... Volunteering would enhance well-being because it boosts self-esteem and builds social resources" [10].

Nanotechnology is a scientific activity, and can be innovated by any individual's idea or effort; but its culmination should be dedicated to the 'welfare' of all of mankind. A widespread profound ethic is to be repaired, re-erected, re-innovated strongly and regularly; everyone should dedicate his/her recreative efforts for the wider world, for needy and deprived people of the world. Nobel Laureate Hans Bathe, one of the most senior of the surviving members of the Manhattan Project which built the atom bomb exploded in Hiroshima, called on all scientists to cease and desist from working on creating, developing, improving, and manufacturing nuclear weapons and other weapons of potential mass destruction [6]. The physicist J. Robert Oppenheimer, who led the building of the atomic bomb in 1945 in the United States, said "It is not possible to be a scientist unless you believe that the knowledge of the world, and the power which this gives, is a thing which is of intrinsic value to humanity, and that you are using it to help in the spread of knowledge and are willing to take the consequences."

It is a dangerous argument, conceived deliberately by the capitalistic, exploitative mindset that competition is the key to innovations and development, even in an unequal, disparity-expanding world.

Nanotechnology-Related Equality, Inclusive Development, and Redefining Intellectual Property Rights; An Apprehension

Today, our society tends to consider only one leg — "liberty" — in isolation and ignore the other two legs — "equality" and "fraternity" — even though all three values are integrated and equally essential for holistic sustainable development. And this can be the only answer for our family conflicts, community and communal conflicts, national and regional conflicts and all sorts of competitions, hatred, non-violence, bombs, and wars.

If there is any possibility of creating a low-priced facility by nanotechnology, by modifying the prevailing patent laws; it should preferably go in favor of eradicating the poverty, ignorance, poor health, lack of employment, anxiety, and depression of the poor. In fact, a very strong ethical value system will refine, reconstruct, and enrich the present "individualistic" "profit"-based system in industry and patent-based system in the R&D and laws. Maybe it seems to be a fantasy in this world of hard constructed capitalistic ethos; but it is the only form and way of a future peaceful and joyful reality. This is the only positive way of thinking and having our peaceful existence, our sustainable economic and mental health, and the joyful development of humanity as a whole.

There is one other dimension regarding nanotechnology that does not divide clearly into commercial and military uses; given their potential in the market, it is hard to imagine pursuing them only in national laboratories. The critical issue will be to make the necessary transparency acceptable in a world of proprietary information, presumably by providing new norms of protection for intellectual property [6].

Considering altruism as the basic bio-socio fiber of humanity, the so-called "liberty" oriented "individual" efforts as "patents" are not the efforts of isolation, but these are the culmination of a long legacy of social, communal, co-operative initiatives. Therefore, some substantial parts of the individual or institutional patents should be distributed to research institutions for the further systematic growth of technologies. Thus, we should reframe the intellectual property rights and "patents" considering a new sharing system.

Hence, One Can Conclude That

Nanotechnology, the manipulation of matter on an atomic and molecular scale, presents a host of ethical considerations as it continues to advance. The intersection of nanotechnology and ethics raises important questions regarding safety, privacy, equity, and the very nature of what it means to manipulate materials at such a small scale. Let's systematically explore some key points in the modern interface of nanotechnology and ethics:

(i) *Impact on the Environment:* Nanowaste disposal presents challenges as tit is not easily degraded and can potentially accumulate in the environment. Ethical considerations include how to responsibly manage and dispose of NPs to minimize harm. Their risk assessment and understanding long-term environmental impact of NPs is crucial. Ethical questions arise concerning who should bear the burden of assessing and mitigating these risks.

(ii) *Health and Safety:* This includes the safety of workers involved in manufacturing and handling nanomaterials; they may face unknown risks. Ethical guidelines are needed to protect workers and ensure they are aware of potential hazards. Moreover, NPs are increasingly being used in consumer goods like cosmetics and clothing. There are debates about whether manufacturers should disclose the presence of nanomaterials and how to ensure consumer safety.

(iii) *Privacy and Surveillance:* Nano-sized sensors could potentially be used for surveillance purposes, raising concerns about privacy and individual freedoms. Ethical guidelines are needed to regulate the use of such technologies.

(iv) *Equity and Access:* There are concerns about the equitable distribution of nanotechnology benefits. Developing countries might not have the resources or infrastructure to access these technologies, leading to potential global disparities. That may be due to cost and accessibility: Ethical questions arise around the affordability and accessibility of nanotechnology-based treatments and products. Will these advancements widen the gap between the wealthy and the disadvantaged?

(v) *Military and Security:* Nanotechnology is being used to develop highly advanced weaponry, raising ethical dilemmas about the morality of creating such destructive capabilities.

(vi) *Ethical Guidelines and Governance:* Governments and international bodies are tasked with creating regulations to ensure the ethical development and use of nanotechnology. Balancing innovation with safety and ethical considerations is a significant challenge. There are calls for transparency in research and development processes involving nanotechnology. Ethical frameworks should promote openness about potential risks and benefits. Ethical decision-making should involve input from various stakeholders, including scientists, policymakers, industry representatives, and the public.

(vii) *Human Enhancement:* Nanotechnology holds promise for enhancing human health and capabilities. Ethical debates surround issues like cognitive enhancement, longevity, and genetic manipulation using nanoscale technologies.

(viii) *Cultural and Societal Impacts:* As nanotechnology becomes more integrated into society, it may challenge traditional values and beliefs. Ethical discussions are needed to navigate these changes thoughtfully. With the advancement in artificial intelligence and nanotechnology the combination of these two will raise complex ethical questions, especially concerning autonomous nanobots and their decision-making capabilities.

Hence, in navigating the modern interface of nanotechnology and ethics, interdisciplinary collaboration is crucial. Scientists, ethicists, policymakers, and the public must engage in ongoing dialogue to ensure that nanotechnological advancements align with ethical principles and contribute positively to society while minimizing risks and harm.

Yugantar Prakash
Department of Psychology & Brain Sciences, Dartmouth College, New Hampshire, USA

References

1. Wang, S., Fu, B., Wei, Y., Lyle, C., Ecosystem services management: An integrated approach. *Curr. Opin. Environ. Sustain.*, 5, 1, 11–15, 2013.
2. Jacobstein, N., Draft version, Foresight Guidelines for Responsible Nanotechnology Development, p. 1, 2006.
3. Madhuri, S., *Advances in Nanomaterials & Applications: The History of Nanotechnology from Pre-historic to Modern Times*, Wiley, USA, 2019.
4. Andrew, C., The Ethics of Nanotechnology, Markkula Center of Ethics, Santa Clara University, 2002.
5. Gunjan, M., Rohayah, B.M., Turani, T., Nanotechnology a Boon or a Curse: A Review. *Open Access J. Toxicol.*, 1, 3, 001–007, 2017.
6. Joy, B., Why the future doesn't need us, issue 8.04, WIRED, April 2000.
7. Madhuri, S., Are we ignoring the considerations for the dark-side of nanotechnology? *J. Nanomed. Res.*, 3, 2, 00051, DOI: 10.15406/jnmr.2016.03.00051.
8. Bainbridge, W.S., Social and ethical implications of nanotechnology, in: *Springer Handbook of Nanotechnology*, B. Bhushan, (Ed.), Springer, Berlin, Heidelberg, 2004, https://doi.org/10.1007/3-540-29838-X_38.
9. Batson, C.D., *Altruism in Humans*, Oxford University Press, New York, 2011.
10. Musick, M.A. and Wilson, J., Volunteerism and depression: The role of psychological and social resources in different age groups. *Soc. Sci. Med.*, 56, 2, 259–269, 2003, doi: 10.1016/s0277-9536(02)00025-4.

Index

Also of Interest

Check out the other titles in the "Advances in Nano-Technology & Applications" series

Risk, Safety, and Ethical Concerns in Nanotechnology
By Madhuri Sharon
Published 2025. ISBN 978-1-119-64038-7

Nanoparticles for Therapeutic Applications
By Madhuri Sharon
Published 2022. ISBN 978-1-119-76230-0

Carbon Nanofibers
Edited by Madhuri Sharon and Maheshwar Sharon
Published 2021. ISBN 978-1-119-76881-4

Nanotechnology in the Defense Industry: Advances, Innovation, and Practical Applications
By Madhuri Sharon, Angelica Silvestre, Lopez Rodriguez, Chetna Sharon and Pio Sifuentes Gallardo
Published 2019. ISBN 978-1-119-46012-1

The History of Nanotechnology: From Prehistoric to Modern Times
Edited by Madhuri Sharon
Published 2019. ISBN 978-1-119-46008-4

Fundamentals of Electrocatalyst Materials and Interfacial Characterization Energy Producing Devices and Environmental Protection
By Nicolas Alonso-Vante
Published 2019. ISBN 978-1-119-46007-7

Behaviors and Persistence of Nanomaterials in Biomedical Applications
By Domenico Cassano and Valerio Voliani
Published 2018. ISBN 978-1-119-41827-6

Carbon Dots as Theranostic Agents
By Madhuri Sharon and Ashmi Mewada
Published 2018. ISBN 978-1-119-46010-7

www.scrivenerpublishing.com

Printed and bound by CPI Group (UK) Ltd, Croydon, CR0 4YY

17/09/2026

14964246-0003